D0968910

GEOLOGICAL ASPECTS OF ACID DEPOSITION

ACID PRECIPITATION SERIES

John I. Teasley, Series Editor

VOLUME 1—Meteorological Aspects of Acid Rain
Edited by Chandrakant M. Bhumralkar

VOLUME 2—Chemistry of Particles, Fogs and Rain
Edited by Jack L. Durham

VOLUME 3—SO_2, NO and NO_2 Oxidation Mechanisms: Atmospheric Considerations
Edited by Jack G. Calvert

VOLUME 4—Deposition Both Wet and Dry
Edited by Bruce B. Hicks

VOLUME 5—Direct and Indirect Effects of Acidic Deposition on Vegetation
Edited by Rick A. Linthurst

VOLUME 6—Early Biotic Responses to Advancing Lake Acidification
Edited by George R. Hendrey

VOLUME 7—Geological Aspects of Acid Deposition
Edited by Owen P. Bricker

VOLUME 8—Economic Perspectives on Acid Deposition Control
Edited by Thomas D. Crocker

VOLUME 9—Modeling of Total Acid Precipitation Impacts
Edited by Jerald L. Schnoor

GEOLOGICAL ASPECTS OF ACID DEPOSITION

Edited by Owen P. Bricker

ACID PRECIPITATION SERIES—Volume 7
John I. Teasley, Series Editor

BUTTERWORTH PUBLISHERS
Boston • London
Sydney • Wellington • Durban • Toronto

An Ann Arbor Science Book

Ann Arbor Science is an imprint of Butterworth Publishers.

Library of Congress Cataloging in Publication Data
Main entry under title:

Geological aspects of acid deposition.

(Acid precipitation series; v. 7)
"An Ann Arbor science book."
Papers presented at a symposium on acid deposition, at the annual meeting of the American Chemical Society, Las Vegas, Nev., Mar. 28-Apr. 3, 1982.
Includes index.
1. Beochemistry. 2. Acid precipitation (Meteorology) I. Bricker, Owen P., 1936– . II. American Chemical Society. III. Series.
QE515.G525 1983 551.9 83–18837

ISBN 0–250–40572–5

Butterworth Publishers
80 Montvale Avenue
Stoneham, MA 02180

10 9 8 7 6 5 4 3 2 1

Printed in the United States of America

CONTENTS

THE EDITORS

JOHN I. TEASLEY, SERIES EDITOR

John I. Teasley was a career employee of the U.S. EPA and several of its predecessor organizations. His background, both academically and throughout his professional employment, was that of an analytical chemist. His duties included research functions as well as serving in line management.

He was quite extensively involved in EPA's acid precipitation program from its inception until his retirement in July 1983. He continues to remain active in the ACS's Division of Environmental Chemistry and currently serves as the Councilor for the Lake Superior Section.

OWEN P. BRICKER, VOLUME EDITOR

Owen P. Bricker is a geochemist with the U.S. Geological Survey, Water Resources Division. He holds a Ph.D. in Geology from Harvard University, MS in Geology from Lehigh University, and BS in Geology from Franklin and Marshall College.

Dr. Bricker's research interests include the chemistry of natural waters, mineral-water interactions, the geochemistry of watershed systems and estuarine and marine geochemistry. He has extensive publications in these areas in professional journals.

THE CONTRIBUTORS

Joan P. Baker
School of Forestry and Environmental Studies
Duke University
Durham, North Carolina

William D. Bischoff
Department of Geological Sciences
Northwestern University
Evanston, Illinois

James J. Bisogni
Department of Environmental Engineering
Cornell University
Ithaca, New York

David L. Correll
Chesapeake Bay Center for Environmental Studies
Smithsonian Institution
Edgewater, Maryland

D.W. Cowell
Lands Directorate
Environment Canada
Burlington, Ontario, Canada

Charles T. Driscoll
Department of Civil Engineering
Syracuse University
Syracuse, New York

Nancy M. Goff
Chesapeake Bay Center for Environmental Studies
Smithsonian Institution
Edgewater, Maryland

Noye M. Johnson
Earth Sciences Department
Dartmouth College
Hanover, New Hampshire

J.S. Kahl
Department of Geological Sciences
University of Maine at Orono
Orono, Maine

A.E. Lucas
Lands Directorate
Environment Canada
Burlington, Ontario, Canada

Fred T. Mackenzie
Oceanography and Hawaii Institute of Geophysics
University of Hawaii
Honolulu, Hawaii

S.A. Norton
Department of Geological Sciences
University of Maine at Orono
Orono, Maine

Virginia L. Paterson
Department of Geological Sciences
Northwestern University
Evanston, Illinois

William T. Peterjohn
Chesapeake Bay Center for
Environmental Studies
Smithsonian Institution
Edgewater, Maryland

Carl L. Schofield
Natural Resources Department
Cornell University
Ithaca, New York

J.S. Williams
Department of Environmental
Protection
Augusta, Maine

SERIES PREFACE

These volumes are a result of a symposium on Acid Precipitation held in conjunction with the American Chemical Society's Las Vegas meeting, held in the Spring of 1982. The symposium was organized along nine thematic areas including meteorology, chemistry of particles, fogs and rain, oxidation of SO_2, deposition both wet and dry, terrestrial effects, aquatic effects, geochemistry of acid rain, economics, and predictive modeling.

The thematic areas were planned and conducted by expert investigators in each of the particular disciplines. The investigators were Chandrakant Bhumralker, meteorology; Jack Durham, chemistry of particles, fogs and rain; Jack Calvert, oxidation of SO_2; Bruce Hicks, deposition; Rick Linthurst, terrestrial effects; George Hendrey, aquatic effects; Owen Bricker, geochemistry; Thomas Crocker, economics; and Jerald Schnoor, modeling.

The symposium was designed to present findings on atmospheric movement of the precursors of acid precipitation; atmospheric chemistry and precipitation effects including aquatic, terrestrial and geochemical; the economics involved with "acid rain" and finally mathematical modeling in order that future problems may be predicted.

As one studies this series it becomes readily evident that the objectives of the symposium were indeed met and that the series will serve as a ready reference in the field of acid precipitation.

I wish to acknowledge the efforts and dedication of the volume editors for the superb job they did in bringing this series to fruition. Working with people such as these made all the time and energy spent well worth while.

I further wish to thank the American Chemical Society, The Division of Environmental Chemistry and the Council Committee of Environmental Improvement for providing the facilities for the symposium from which this series evolved.

Finally, I wish to express my gratitude to the individual chapter authors without whom the series would never have been possible.

John I. Teasley

PREFACE TO VOLUME 7

In recent years "acid rain" has become a common term in the scientific literature and the popular press. Mention of the term conjures up images of fishless lakes, dying forests and other types of damage to biological systems. Thus far, emphasis in the acid rain arena has centered on impacts on the biosphere, hydrosphere and atmosphere. The chapters in this volume deal primarily with the lithosphere. Too often we forget that the lithosphere (bedrock, sediment, soil) forms the containment for lakes, streams and groundwaters—the waters whose ultimate source has been atmospheric deposition. These waters flow on and circulate through the materials comprising the lithosphere, react with them, and both the waters and the solids are changed chemically through these interactions. The general pattern of rain falling on the earth and reacting with the materials of the lithosphere (the weathering reactions so familiar to every beginning geology student) began soon after the earth was formed and has continued to the present. Anthropogenic additions to the natural acidic components of the atmosphere have increased since the time of the industrial revolution until they now rival or exceed those of the natural system. The severity of the environmental perturbations caused by these anthropogenic additions to the atmosphere has become a hotly debated topic in scientific forums and in the political arena.

The six chapters in this book address various aspects of the acid deposition phenomenon from a geological perspective. It is hoped that the geological approach will be useful in bringing the problem more clearly into focus and may shed light on the geochemical processes that modify the chemical composition of acid deposition after it encounters and reacts with the materials of the lithosphere.

I would like to express my appreciation to the American Chemical Society for hosting the symposium on acid deposition at its annual meeting in Las Vegas, Nevada, March 28–April 2, 1982, when these papers were presented; to Dr. John Teasley, who organized the symposium and is the series editor for the resultant publications; to the reviewers who made significant contributions to improving the volume; and most of all to the authors of the contributions appearing in this volume.

Owen P. Bricker

CHAPTER 1

Geochemical Mass Balance for Sulfur- and Nitrogen-Bearing Acid Components: Eastern United States

William D. Bischoff
Virginia L. Paterson
Fred T. Mackenzie

The impact on a geographical region of SO_2 and nitrogen oxides (NO_x) emissions to the atmosphere because of man's activities (e.g., burning of fossil fuels and smelting of sulfide ores) usually has not been considered in terms of a regional geochemical mass balance model. Mass balance models, however, have been employed extensively on a global scale [1–4]. The models evaluate reservoir sizes, processes and fluxes associated with the transfer of a substance within a system of interest. The models may be steady- or transient-state, and include assessment of historical (geologic), present and future data and processes.

In this chapter a geochemical mass balance model is applied to constituents of acid precipitation (H^+, NO_3^- and SO_4^{2-}) to evaluate the impact of acid precipitation on the eastern United States. The eastern United States is defined as all states east of the Mississippi River, excluding small portions of Minnesota and Louisiana. The model is composed of an atmospheric reservoir and a terrestrial reservoir and involves fluxes of acid rain constituents between these reservoirs or out of the entire system (see Figure 6). The model is restricted to anthropogenic fluxes and, at present, reservoirs are considered to be in steady state. The mass balance of the atmospheric reservoir is maintained by influx of SO_2 and NO_x from anthropogenic emissions and outflux of related species by precipitation, dry deposition and atmospheric transport (wind). Accumulation of sulfur- and nitrogen-bearing acid components in the atmosphere is shown to be small. Influx of sulfur- and nitrogen-bearing constituents by atmospheric transport from outside the system (western United States, Canada or the Atlantic Ocean) is considered insignificant to other fluxes. Influxes of precipitation and dry deposition, and outflux by streams maintain the mass balance of the terrestrial reservoir. The material balance equations are, for the atmospheric reservoir:

$$\frac{dC_{H^+}}{dt} = E - P - D - W - O$$

and for the terrestrial reservoir:

$$\frac{dC_{H^+}}{dt} = P + D - S = O$$

where dC_{H^+}/dt is rate of change of hydrogen ion concentration; E, P, D, W and S denote fluxes of H^+ related to anthropogenic emissions, precipitation, dry deposition, wind and streams, respectively, of sulfur- and nitrogen-bearing constituents.

The following sections discuss the procedures used to derive the flux estimates. These estimates were obtained by direct calculation from concentration and total flow (river discharge, rainfall) data, or by model considerations commonly involving difference calculations. Because of the large amount of data initially available by state, the state was the geographical entity used to obtain most of the flux estimates, usually by summation and averaging, for the entire eastern United States. In conclusion, the model and its strengths and weaknesses are discussed, the entire mass balance for sulfur- and nitrogen-bearing acid constituents in the eastern United States presented, and some comments on the regional stream cation balance and future research are made.

DIRECT FLUX ESTIMATES

Fluxes for the mass balance model were estimated directly from available data or by difference calculations. All fluxes are reported in units of 10^{10} mol-y^{-1} of S or N. Emissions are actually in the form of the chemical species SO_2 and NO_x. Wet deposition species are SO_4^{2-} and NO_3^-, and dry deposition and atmospheric transport fluxes include the species SO_2, SO_3, SO_4^{2-}, NO, NO_2, and NO_3^-.

Emissions

Annual estimates of SO_2 and NO_x emissions by state were obtained from reports of the U.S. Environmental Protection Agency (EPA) for the period 1972–1977 [5–10]. The EPA and affiliated national and state agencies monitor atmospheric pollution sources, including fuel combustion for transportation and energy needs (residential, industrial, governmental and commercial), industrial processes (chemical manufacturing, ore smelting, etc.), incineration of solid wastes, and accidental and managed agricultural, structural and forest burnings. These estimates do not include agricultural emissions of S and N related species. Details of calculation methods are described in EPA reports [5–10]. These emission estimates were averaged by state over the six-year period and summed to give total annual emission estimates of SO_2 and NO_x for the eastern United States of 35.6×10^{10} and 27.7×10^{10} mol, respectively. EPA emission estimates are reported as weights. We converted the EPA values to moles assuming all NO_x was NO_2. The resulting flux of NO_x, therefore, is a minimum. Variation in annual emission estimates indicates that the calculated fluxes are within a ±15% range of actual values for any year in the six-year period. Yearly state averages are listed in Tables I to IV.

Table I. Area, Average Annual Precipitation, Runoff and Emissions of SO_2 and NO_x for States Within the Eastern United States

	Area (km^2)	*Precipitation* (10^{11} L)	*Runoff* (10^{11} L)	SO_2 *Emissions* (10^3 t)	NO_x *Emissions* (10^3 t)
Alabama	130,400	1,814	852	934.3	413.7
Connecticut	12,492	149	119	107.8	183.6
Delaware	5,084	59	(with MD)	173.3	56.2
Florida	138,823	1,876	515	883.9	654.2
Georgia	148,858	1,828	671	563.5	414.2
Illinois	143,279	1,362	451	2,201.1	1,161.1
Indiana	92,794	933	417	1,899.2	924.6
Kentucky	102,328	1,190	772	1,409.3	499.7
Maine	83,782	884	658	137.1	81.4
Maryland and the District of Columbia	25,384	281	159	375.5	316.0
Massachusetts	23,966	270	149	375.6	345.7
Michigan	145,625	1,157	402	1,289.4	981.4
Mississippi	121,753	1,666	972	141.2	206.4
New Hampshire	23,164	251	(with VT)	98.2	72.6
New Jersey	19,315	226	102	362.7	478.8
New York	122,748	1,193	541	915.3	884.5
North Carolina	125,812	1,550	720	540.9	506.0
Ohio	105,194	997	532	3,186.0	1,169.3
Pennsylvania	115,464	1,225	829	3,703.4	1,323.8
Rhode Island	2,697	29	5	36.0	44.5
South Carolina	77,648	891	298	244.3	300.4
Tennessee	105,953	1,356	752	1,172.2	480.4
Vermont	23,774	242	346	10.8	24.3
Virginia	126,712	1,378	543	453.8	383.6
West Virginia	61,846	702	375	956.8	402.7
Wisconsin	139,651	1,093	500	645.6	424.7
Total	2,224,546	24,602	11,680	22,817.2	12,733.8

Table II. NADP Average Precipitation pH and Concentrations of H^+, SO_4^{2-} and NO_3^- for October 1979 to September 1980[a]

NADP Site	*State*	*Quarters Sampled*	*pH*	*H^+ (μM)*	*SO_4^{2-} (μM)*	*NO_3^- (μM)*	*SO_2 (μg-m^{-3})*
Fayetteville	AK	2	4.84	14.5	19.2	19.5	60
Austin-Cary Forest	FL	4	4.76	17.6	17.0	10.5	70
Bradford Forest	FL	4	4.76	17.4	12.7	10.1	
Everglades Natl Park	FL	2	4.88	13.1	11.1	10.6	40
Georgia Station	GA	4	4.49	32.7	21.3	14.5	
Bondville	IL	4	4.19	65.1	36.2	30.5	
Argonne	IL	3	4.20	63.7	41.4	38.1	
SIU	IL	4	4.34	45.3	30.8	25.5	
Salem	IL	2	4.18	65.7	35.4	28.2	
Dixon Springs Ag. Ctr.	IL	4	4.27	53.3	39.9	25.1	
Indiana Dunes Nat. Lakeshore	IN	1	4.09	81.3	47.3	36.4	340
Caribou	ME	2	4.45	35.3	23.4	18.0	
Greenville Stn.	ME	4	4.33	46.2	25.7	22.1	200
Univ. of Michigan Biological Stn.	MI	4	4.34	45.8	29.9	32.6	
Isle Royale Nat. Pk.	MI	1	4.65	27.1	21.1	19.9	
Kellogg Biol. Stn.	MI	4	4.32	47.6	37.6	36.0	
Wellston	MI	4	4.30	50.1	31.4	37.2	
Marcell Exp. Forest	MN	4	4.92	11.9	18.0	22.9	30
Lamberton	MN	4	4.93	11.9	26.3	31.3	
Meridian	MS	2	4.70	20.0	12.9	13.3	
Hubbard Brook	NH	4	4.29	50.8	25.3	27.1	
Aurora Research Farm	NY	4	4.13	73.6	40.4	36.9	
Chautauqua	NY	2	4.00	100.4	51.4	38.3	500
Knobit	NY	3	4.15	70.6	26.0	28.3	
Huntington Wildlife	NY	4	4.28	53.0	39.7	38.2	
Stillwell Lake, West Point	NY	4	4.20	62.7	30.0	26.2	
Bennett Bridge	NY	2	3.69	202.5	69.6	51.9	800
Jasper	NY	3	4.16	69.7	32.5	31.7	
Lewiston	NC	4	4.46	34.2	22.1	19.4	180
Coweeta	NC	4	4.53	29.5	17.7	12.6	180
Research Triangle Pk	NC	2	4.49	32.1	19.9	12.6	180
Piedmont Res. Stn.	NC	4	4.47	33.8	29.8	21.6	180
Clinton Crops Res Stn	NC	4	4.54	28.8	17.8	14.5	180
Finley	NC	4	4.54	28.6	17.7	14.8	180
Delaware	OH	4	4.13	73.4	40.2	34.4	300
Caldwell	OH	4	4.10	78.9	42.9	31.5	
Wooster	OH	4	4.17	67.6	39.6	31.9	300

[a] NADP stations with less than four quarters sampled were established during the sampling period. Estimated atmospheric second maximum SO_2 concentrations are from the EPA [12].

Table II. (*Continued*)

NADP Site	*State*	*Quarters Sampled*	*pH*	*H^+ (μM)*	*SO_4^{2-} (μM)*	*NO_3^- (μM)*	*SO_2 (μg-m^{-3})*
Kane Exp. For.	PA	4	4.15	71.6	35.1	31.8	
Leading Ridge	PA	4	4.15	70.2	33.9	31.8	
Clemson	SC	4	4.50	31.6	19.2	14.2	180
Walker Branch Wtrsd.	TN	4	4.17	68.1	33.6	22.2	275
Horton's Station	VA	4	4.37	42.2	27.9	19.5	200
Parsons	WV	4	4.19	64.1	35.9	31.7	
Trout Lake	WI	3	4.68	20.8	18.6	23.0	130
Spooner	WI	2	5.30	49.4	21.3	27.2	

Annual Precipitation and Wet Deposition

To obtain an estimate of average annual precipitation in the eastern United States, equal area precipitation maps averaged for a 25-y period were used [11]. A square grid of 256-km² segments was constructed and rainfall in each square was deter-

Table III. Drainage Area, Annual Discharge and Flux of SO_4^{2-} and $NO_2^- + NO_3^-$ for 16 Principal Rivers and 140 Total Rivers in the Eastern United States (Data Are Primarily from Water Year 1978)

River, State	*Discharge (10^{10} L)*	*Drainage Area (km²)*	*SO_4^{2-} Flux (10^{10} mol)*	*$NO_2^- + NO_3^-$ Flux (10^{10} mol)*
Ohio, IL	30,302	526,029	14.99	2.38
St. Marys, MI	5,448	209,531	0.15	0.11
Susquehanna, MD	3,350	70,189	1.28	0.29
Alabama, AL	3,008	55,685	0.25	0.05
Illinois, IL	2,558	73,981	2.18	0.77
Delaware, NJ	1,479	17,560	0.31	0.09
Potomac, DC	1,407	29,940	0.51	0.11
Penobscot, ME	1,251	17,275	0.10	0.06
Savannah, GA	1,204	25,512	0.07	0.03
Yazoo, MS	1,165	19,814	0.09	0.03
Pee Dee, SC	1,103	22,870	0.10	0.05
Roanoke, NC	1,085	21,782	0.09	0.02
Hudson, NY	1,008	20,953	0.19	0.04
James, VA	855	17,503	0.14	0.03
Suwanee, FL	748	20,409	0.10	0.02
Connecticut, CT	183	28,184	0.03	0.01
Total	56,154	1,172,217	20.6	4.1
140 Rivers Total	74,536		24.8	4.6

Table IV. Flux of SO_4^{2-} and NO_3^- of 12 Principal Rivers Analyzed before 1924 (Discharge Listed is from Water Year 1978) [18]

River, State	*Discharge (10^{10} L)*	*SO_4^{2-} Flux (10^{10} mol)*	*NO_3^- Flux (10^{10} mol)*
Susquehanna, PA	3,953	1.28	0.22
Alabama, AL	3,008	0.33	0.04
Illinois, IL	2,625	1.15	0.18
Hudson, NY	1,234	0.21	0.02
Connecticut, NH	1,087	0.08	
Savannah, GA	1,081	0.07	0.01
Penobscot, ME	1,058	0.07	0.00
Delaware, NJ	1,049	0.13	0.02
Potomac, DC	1,026	0.21	0.06
Pee Dee, SC	866	0.04	0.01
Roanoke, VA	725	0.03	0.01
James, VA	669	0.05	0.00
Total	18,381	3.60	0.60

mined. Total average annual rainfall in the eastern United States is 2.46×10^{15} L. Comparable values were determined on a state-by-state basis (Tables I to IV).

To estimate wet deposition of H^+, SO_4^{2-} and NO_3^-, isoplethic maps of pH and SO_4^{2-} and NO_3^- concentrations were constructed (Figures 1 to 3). Contour lines principally were constrained by the geographical distribution of sampling stations of the National Atmospheric Deposition Program (NADP). In areas where NADP stations have not been established, however, contour line configurations were governed by the pattern of contours outside the area of interest and by EPA emission density maps [12], and the Likens et al. [13] map of pH isopleths.

Since the NADP started in late 1978, the number of sampling stations has increased steadily throughout the United States from 22 in late 1978 to 74 in the third quarter of 1980. Most NADP stations collect precipitation continuously, and conduct analyses weekly. For each station in the eastern United States, weekly analyses of H^+, SO_4^{2-} and NO_3^- were averaged over the period from October 1979 through September 1980 [14,15]. The volume-weighted, annual average concentrations for each station are listed in Tables I to IV. By overlaying the isoplethic maps onto the rainfall grid, the total annual wet depositions of SO_4^{2-}, NO_3^- and H^+ for the eastern United States were calculated as 6.7×10^{10}, 5.4×10^{10} and 12.1×10^{10} mol, respectively.

Wet deposition also was calculated by state by choosing appropriate concentrations of SO_4^{2-} and NO_3^- based on nearby NADP stations and state precipitation. These estimates are 6.6×10^{10} and 5.7×10^{10} mol-y^{-1} of SO_4^{2-} and NO_3^-, respectively. One other attempt was made to calculate wet deposition fluxes. Plots of H^+ vs SO_4^{2-} and H^+ vs NO_3^- concentrations for the NADP stations were constructed

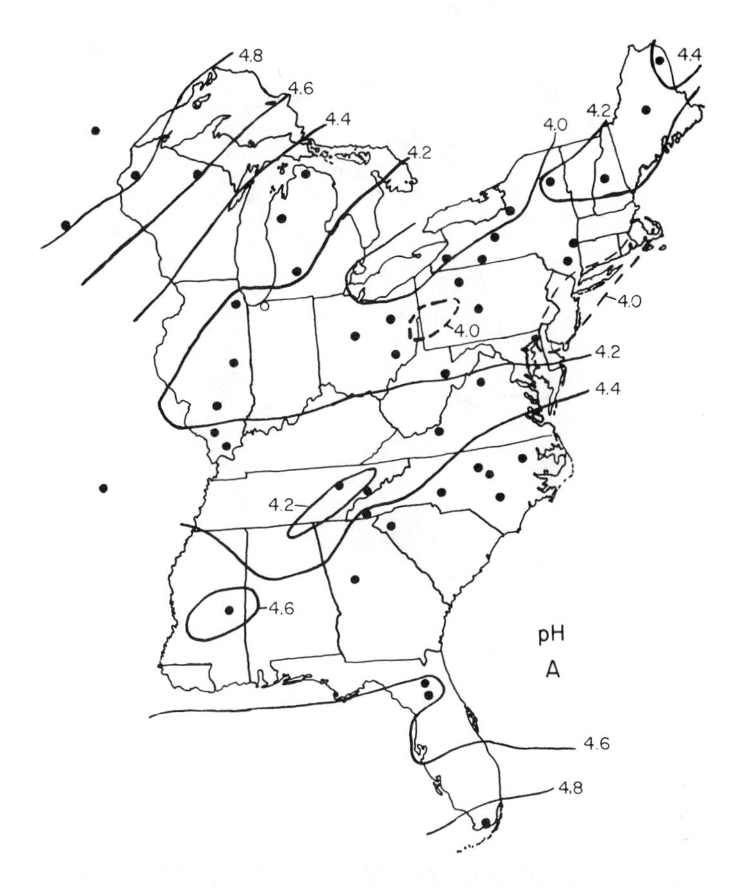

Figure 1. Isoplethic map of annual precipitation pH in the eastern United States. Black circles are NADP stations. Dashed isopleths are inferred on the basis of EPA emission density maps [27].

and subjected to least squares analysis (Figure 4). The average concentration of H^+ for the eastern United States (49.3 μM) was then used to calculate average concentrations of SO_4^{2-} and NO_3^-. The wet deposition of SO_4^{2-} and NO_3^-, by this method, is estimated as 7.2×10^{10} and 6.3×10^{10} mol-y^{-1}, respectively.

We adopt as the best wet deposition values those derived by using isoplethic maps, although they are probably subject to an error of $\pm 1 \times 10^{10}$ mol-y^{-1}. These values apply to a one-year period, and probably do not represent a multiyear mean.

Runoff

The runoff of SO_4^{2-} and NO_3^- from the eastern United States to the sea includes natural products from rock weathering as well as SO_4^{2-} and NO_3^- from anthropogenic

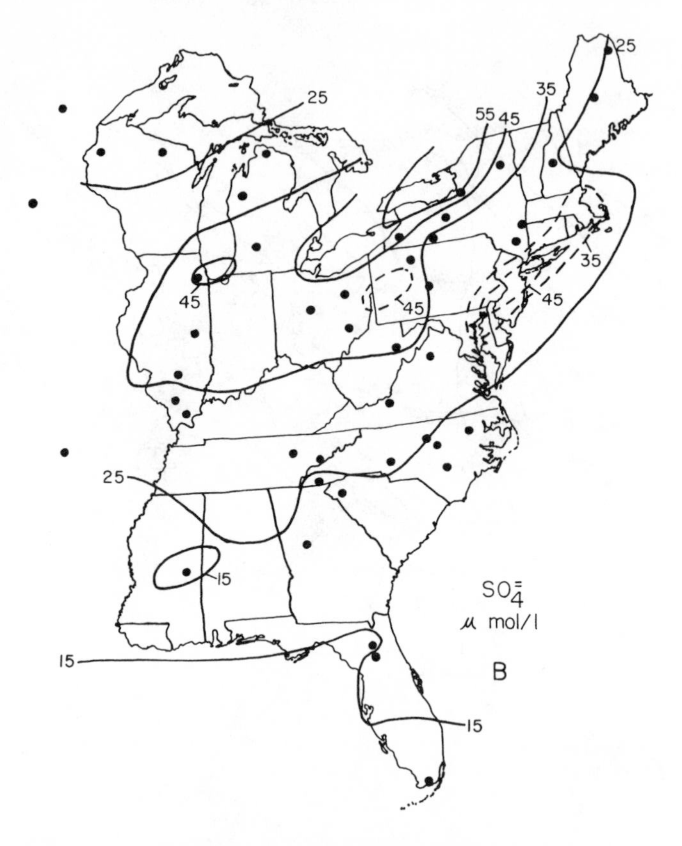

Figure 2. Isoplethic map of annual precipitation SO_4^{2-} concentration in the eastern United States. Black circles are NADP stations. Dashed isopleths are inferred on the basis of EPA emission density maps [27].

sources. U.S. Geological Survey water resources data for all states in the eastern United States, as reported in the volumes of the U.S. Geological Survey Water Data Reports (mainly for 1978, but also 1977, 1979 and 1980), were used to compute SO_4^{2-} and NO_3^-. Most of the data are for the water year 1978 (October 1, 1977 through September 30, 1978), but where data for that year were not available, data from other water years ranging from 1976 to 1980 were used.

Runoff from the entire eastern United States was determined from an average ratio of runoff to precipitation. To calculate the runoff-to-precipitation ratio for each state, the total annual water discharge of major river basins in the state was divided by the drainage area of the basins. An average value of runoff per square mile was then multiplied by the state area to give total runoff for the state. Runoff-to-precipitation ratios were calculated for each state (Tables I to IV), weighted according to state area, and averaged over the entire eastern United States. The average ratio of runoff to precipitation was computed as 0.47, giving a total runoff value of 1.16×10^{15} L.

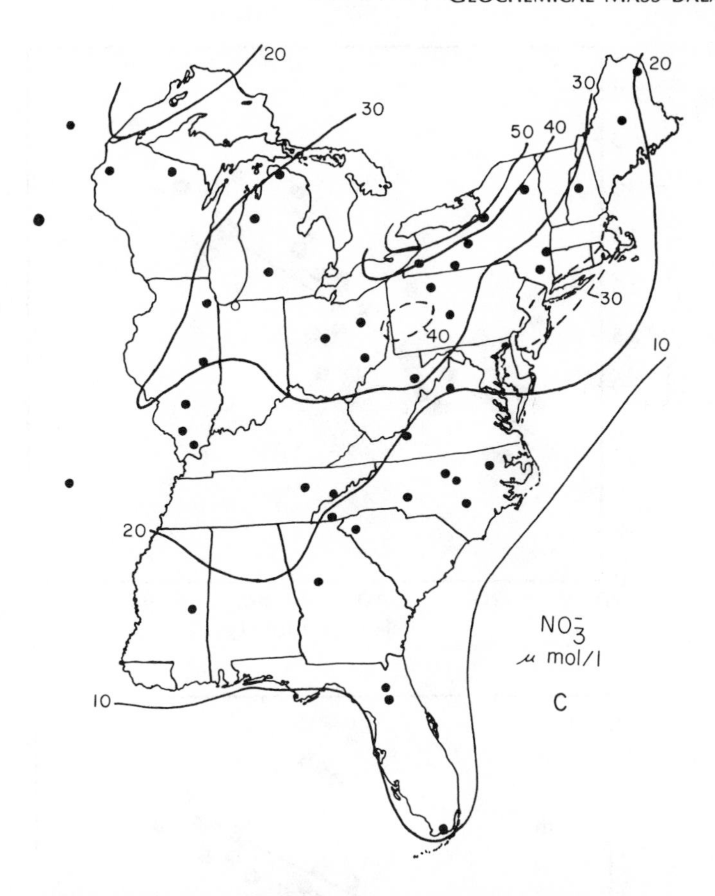

Figure 3. Isoplethic map of annual precipitation NO_3^- concentration in the eastern United States. Black circles are NADP stations. Dashed isopleths are inferred on the basis of EPA emission density maps [27].

Runoff fluxes of SO_4^{2-} and NO_3^- were determined from data of 140 rivers. All of these rivers flow to the perimeter of the eastern United States, directly to the sea or indirectly to the sea via the Mississippi and St. Lawrence Rivers. No rivers were considered that eventually flow into another river inside the region. Total annual discharge of the 140 rivers is 7.45×10^{14} L. These rivers have an average SO_4^{2-} concentration of 333 μM and NO_3^- concentration of 61.7 μM. Adjusting for the difference between the discharge of the 140 rivers and runoff from the entire eastern United States, the annual runoff of SO_4^{2-} and NO_3^- is calculated to be 38.6×10^{10} and 7.2×10^{10} mol, respectively. A similar calculation for 16 principal rivers results in annual runoff fluxes of 42.6×10^{10} mol SO_4^{2-} and 8.5×10^{10} mol NO_3^- (Tables I to IV). We have adopted values of $40 \pm 3 \times 10^{10}$ mol SO_4^{2-} per year and $7.8 \pm 1 \times 10^{10}$ mol NO_3^- per year.

To obtain some idea of the amount of SO_4^{2-} and NO_3^- of anthropogenic origin in eastern United States rivers, we attempted to estimate pristine river loads

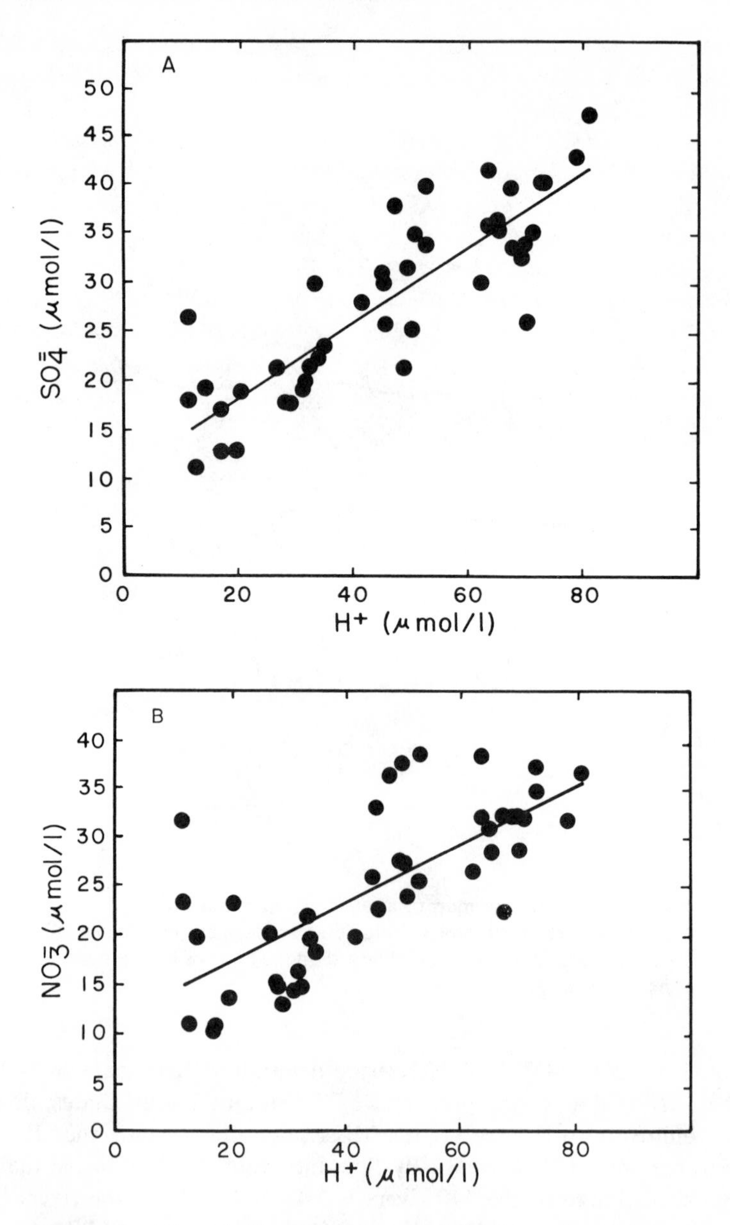

Figure 4. Plots of H^+ vs (A) SO_4^{2-} and (B) NO_3^- averaged annual concentrations in precipitation for NADP data. Least squares equations are $SO_4^{2-} = 0.385(H^+) + 10.3$, and $NO_3^- = 0.298(H^+) + 11.3$.

of these constituents. This task is difficult because of the lack of pre-1900 species concentration data for rivers. Also, it is dangerous to use today's total river runoff value for ancient times because of climatic change and consequent changes in the historical pattern of precipitation. Nevertheless, an attempt was made to estimate the proportion of today's SO_4^{2-} and NO_3^- discharge attributable to rock weathering, and by difference, to calculate the anthropogenic contribution of these species to eastern United States rivers.

The river fluxes of SO_4^{2-} and NO_3^- were corrected for products of natural rock weathering from estimates of "pristine" SO_4^{2-} and NO_3^- concentrations. Meybeck [16] calculated pollution-corrected, average runoff concentrations of SO_4^{2-} from different rock terrains. The estimates are 20.8 μM SO_4^{2-} for plutonic and highly metamorphosed rocks and 260 μM for sedimentary rocks. The eastern United States is composed of approximately 80% sedimentary rocks and 20% plutonic or highly metamorphosed rocks (Figure 5). If the source area of runoff is proportional to rock type, 80% of the runoff is from sedimentary rocks and 20% from plutonic or highly metamorphosed rocks. The pristine annual runoff

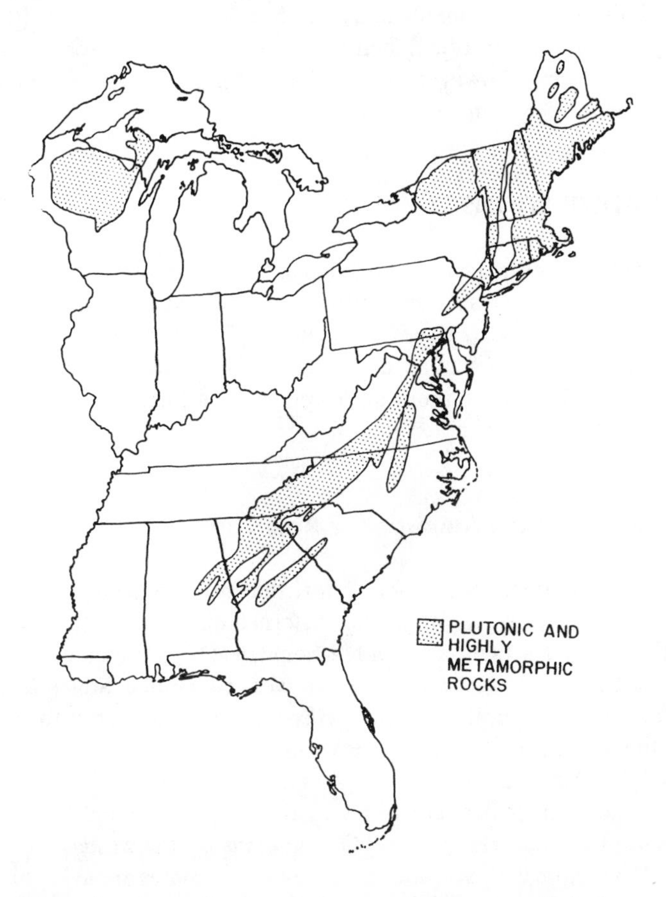

Figure 5. Generalized geologic map of the eastern United States. White areas are sedimentary rocks; dotted areas are plutonic and highly metamorphic rocks. Sedimentary rocks comprise approximately 80% and plutonic and highly metamorphic rocks comprise approximately 20% of the eastern United States land area.

of SO_4^{2-} is calculated to be 24.2 × 10^{10} mol. Wollast [17] calculated the pristine world average runoff of NO_3^-. Although no distinction is made for NO_3^- concentrations from different rock types, the value of 3 μM gives a pristine flux of 0.4 × 10^{10} mol-y^{-1} of NO_3^- from eastern United States rivers to the sea.

Pristine runoff estimates also were calculated on the basis of pre-1924 river analyses of SO_4^{2-} and NO_3^- concentrations reported by Clarke [18] (Tables I to IV). The total discharge of SO_4^{2-} and NO_3^- for these 12 major rivers of the eastern United States, with a total runoff of 1.84 × 10^{14} L-y^{-1} adjusted for today's total runoff, is 22.7 × 10^{10} and 3.8 × 10^{10} mol, respectively.

From the above calculations and considerations, we adopt values of 24 × 10^{10} mol SO_4^{2-} and 1 × 10^{10} mol NO_3^- for the pristine annual runoff of these constituents from the eastern United States to the sea. Using today's discharge of these constituents and their pristine values, by difference, the river fluxes of SO_4^{2-} and NO_3^- related to man's activities are found to be 16 × 10^{10} and 6.8 × 10^{10} mol-y^{-1}, respectively. It is difficult to assess the accuracy of the above calculations. It is rewarding, however, to see that different approaches give reasonably consistent pristine SO_4^{2-} and NO_3^- fluxes.

MODEL CALCULATIONS

Natural inputs of sulfur- and nitrogen-bearing compounds into the atmosphere of the eastern United States are insignificant compared to the magnitudes of anthropogenic fluxes. The model, therefore, primarily deals with these latter processes and fluxes. The following section describes the procedure for calculation of fluxes by differences, and presents minor anthropogenic and natural fluxes, and comparisons with previously determined fluxes.

Emissions and the Atmospheric Reservoir

It can be demonstrated, at least for sulfur, that no significant inputs of emissions enter the atmospheric reservoir of the eastern United States from Canada or the western United States Galloway and Whelpdale [19] estimated that these fluxes for SO_2 are 1.2 × 10^{10} mol-y^{-1} from the western United States and 2.2 × 10^{10} mol-y^{-1} from Canada. Both of these fluxes are small compared to total emissions of SO_2 to the atmosphere from sources within the eastern United States. Emissions of SO_2 in the eastern United States comprise 80% of the total U.S. SO_2 emissions. Because the amount of SO_2 emissions in the western United States is relatively small, it is unlikely that transport of SO_2 emissions for the western U.S. atmosphere across the Mississippi is significant. Although no estimates are available for external fluxes of NO_x, we assume that these also are unimportant. Emissions of NO_x in the western United States, however, comprise 40% of the total U.S. NO_x emissions. Transport of NO_x from the western to the eastern United States is probably more significant than transport of SO_2 emissions.

Biogenic emissions of SO_2 (marine and terrestrial) into the eastern U.S. atmosphere were estimated by Galloway and Whelpdale [19] to be 1.4×10^{10} mol-y^{-1} of SO_2 and are, again, small compared to total emission. Sea aerosol transport of SO_4^{2-} into this region also was found to be insignificant [19]. The total flux of SO_2 from outside the eastern United States and from biogenic sources is considerably less than the 15% range about the mean in emission estimates for SO_2 for the six-year period.

The magnitudes of natural sources of NO_x to the atmosphere in the eastern United States are most difficult to evaluate. Controversy still exists as to the global importance of denitrification processes leading to losses of NO_x from agricultural and natural soils to the atmosphere [20–24]. It is possible, however, to show that the potential maximum flux of NO_x from eastern U.S. agricultural and natural soils is very likely small compared to the anthropogenic nitrogen flux to the atmosphere.

Laboratory measurements on the evolution of nitrogen dioxide from soils, although few, give values in the range $0.09–2.2 \times 10^{-1}$ g-$m^{-2}y^{-1}$ of N [25,26]. At the maximum emission rate, the surface area of the eastern United States could give a natural flux of 6×10^{10} mol NO_x. This natural source of NO_x for the eastern U.S. atmosphere would be about 20% of the anthropogenic emissions flux.

We realize the speculative nature of the above calculations, but on the basis of present knowledge of the significance of soil losses of NO_x, it is likely that anthropogenic emissions of NO_x to the eastern U.S. atmosphere are significantly more important than natural emissions.

A possible sink for atmospheric emissions of SO_2 is its accumulation in the atmospheric reservoir. The short residence time of SO_2 (measured in days) suggests that accumulation is unlikely. A simple calculation documents this assertion. Second maximum 24-h average SO_2 concentrations [27] were plotted against average annual concentrations of H^+ from NADP stations where data overlapped (Figure 6). The average annual concentration of H^+ in the entire eastern United States of 49.3 μM was used to calculate a maximum 24-h gaseous SO_2 concentration. The maximum is 227 μg-m^{-3} of $SO_{2(g)}$ and probably is never achieved over the entire eastern United States. Ambient concentrations of gaseous SO_2 are actually close to 7 μg-m^{-3} [19]. Using the entire troposphere with an average thickness of 15 km, and an eastern U.S. surface area of 2.22×10^6 km^2, the volume of the model atmospheric reservoir is 3.3×10^{16} m^3. Total gaseous SO_2 in this reservoir at *maximum* SO_2 concentration would be 12×10^{10} mol. At the present rate of SO_2 emissions to the eastern U.S. atmosphere owing to man's activities, it would take only four months to accumulate this amount of SO_2 in the atmosphere, provided all the emissions remained there.

The average pH of rainfall over the eastern United States has not changed drastically since 1955, when the minimum pH was about 4.4 [24]; current average minimum pH is about 4.0. In the past, the pH of acid rain primarily was controlled by SO_2, until recently when NO_x also has become important [24,25]. Therefore, it is likely that if gaseous SO_2 has accumulated significantly in the eastern U.S. atmosphere, it accumulated before the mid-1950s. If all the above calculated SO_2 of 12×10^{10} mol did accumulate in the atmosphere over an approximate 25-y period at a constant rate, only 1%-y^{-1} of the average annual anthropogenic SO_2

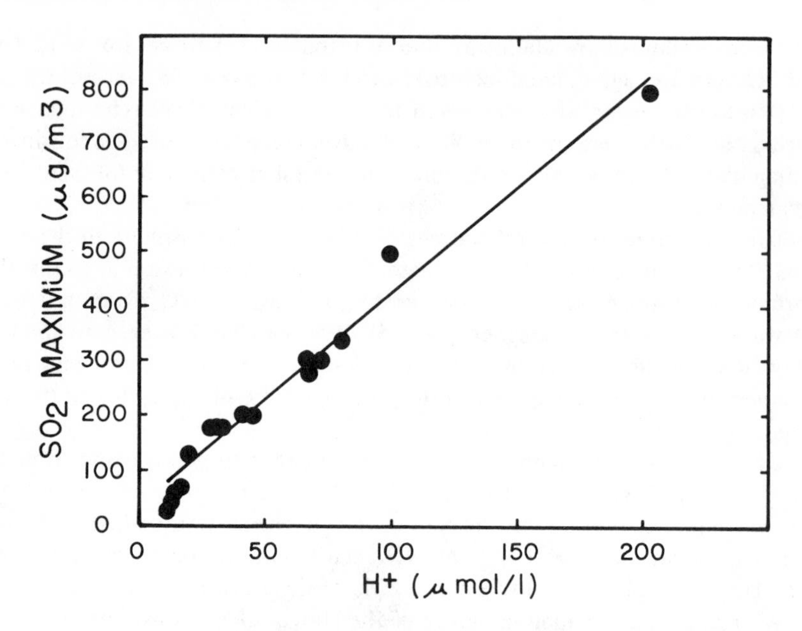

Figure 6. Plot of atmospheric, second maximum 24-h SO_2 concentrations vs NADP averaged annual concentration of H^+ in precipitation. The least squares equation is: $SO_{2(g)} = 3.93(H^+) + 33.2$ μg-m^{-3}. The average annual concentration of H^+ in precipitation of 49.3 μM in the eastern United States gives a second maximum SO_2 concentration of 227 μg-m^{-3}.

emission of 35.6×10^{10} mol would be necessary to account for this accumulation. Whatever the case, it is unlikely that the eastern U.S. atmospheric reservoir is an important sink of SO_2.

The global residence time of NO_x is about one month [23]. This relatively short residence time and considerations similar to those above for SO_2 do not indicate that any significant accumulation of NO_x in the atmosphere occurs on a yearly basis.

Wet Deposition

The annual wet depositions of SO_4^{2-}, NO_3^- and H^+ are 6.7×10^{10}, 5.4×10^{10} and 12.1×10^{10} mol, respectively. Wet deposition is responsible for removal of 19% of the SO_2 emissions and 19.5% of the NO_x emissions. Average annual concentrations of SO_4^{2-}, NO_3^- and H^+ in wet deposition are 27.2, 22.1 and 49.3 μM, respectively.

Maximum contributions to free acidity by HNO_3 (NO_3^-/H^+) and H_2SO_4 ($2SO_4^{2-}/H^+$) are 45 and 100%, respectively. The increasing importance of HNO_3 as a contributor to acid precipitation was shown for the Hubbard Brook Experimental Forest, New Hampshire, by Galloway and Likens [27]. They found that the current maximum contribution of HNO_3 is 31% in the summer and 61% in the winter. The annual value for the eastern United States as a whole falls between these values, demonstrating the importance of HNO_3 as an acid contributor to the entire eastern United States.

Galloway and Whelpdale [19] estimate a wet deposition flux of 7.8×10^{10} mol-y^{-1} of SO_4^{2-} for the eastern United States. They suggest that this value may be high by 30% because of problems associated with wet deposition sampling, such as inefficient collection, sample evaporation and contamination. Much of their wet deposition data were collected on a monthly basis. The NADP samples are collected weekly and analyzed by a uniform procedure, helping to eliminate many of the aforementioned problems.

Dry Deposition

Dry deposition of S (as SO_2, SO_3 and SO_4^{2-}) and N (as NO_x and NO_3^-) was calculated from the difference between the wet deposition fluxes and anthropogenic river fluxes. Dry deposition of S is 9.3×10^{10} mol-y^{-1} and of N, 1.4×10^{10} mol-y^{-1}. Dry deposition is responsible for removal of 26% of the SO_2 emissions and 5% of the NO_x emissions.

It is possible that acid mine drainage may account for some of the anthropogenic river flux of SO_4^{2-}. Our estimate of SO_4^{2-} dry deposition, however, is very near that obtained by Galloway and Whelpdale [19] of 10.3×10^{10} mol S. Their estimate was calculated independently from deposition velocities and regional SO_2 concentrations. The good agreement between these SO_2 values suggests that, whereas acid mine drainage may be important locally [28], it has no significant effect over the entire area of the eastern United States.

Because some NO_3^- in riverwater is derived from fertilizers, the NO_x dry deposition flux, as calculated, is a maximum estimate. Furthermore, denitrification processes may lead to production of gaseous N_2, N_2O and NH_3 in soils. If NO_3^- added to soils of the eastern United States is used in denitrification reactions, resulting in production of N_2, N_2O and NH_3, then the *total* (wet and dry) depositional flux of NO_3^- in acid precipitation could be a maximum net value. Few data are available to quantify the significance of denitrification in returning originally nitrogen-bearing acid constituents to the atmosphere. It can be shown by using the global N_2, N_2O and NH_3 soil denitrification estimates of Söderlund and Svensson [21], adjusted for the land area of the eastern United States, that the denitrification flux easily could be equal to the total depositional flux of nitrogen in acid precipitation. This certainly is not the situation in the eastern United States, but there is a strong need for information regarding this process and related fluxes.

Atmospheric Transport

It has been shown that storage in the atmospheric reservoir is not a significant sink for emissions. Emissions not accounted for by wet or dry deposition must

rn U.S. atmospheric reservoir by wind. Atmospheric transport, there- for 55% of the original SO_2 emissions (19.6×10^{10} mol-y^{-1} of S) he NO_x emissions (20.9×10^{10} mol-y^{-1} of N).

nospheric transport of potential acid precipitation to Canada and the Atlantic has been estimated. The heavy line in Figure 7 divides the region of the eastern United States in which the average annual winds blow toward Canada from the region in which winds blow toward the Atlantic. Emissions in the areas with wind vectors in the direction of Canada account for approximately 52 and 45% of the total SO_2 and NO_x emissions, respectively, in the eastern United States. Assuming that wet and dry deposition are proportional to emissions in the area overall, these percentages can be used to calculate fluxes to Canada and the Atlantic from the overall atmospheric transport flux. This crude approximation leads to estimates of 10.2×10^{10} mol-y^{-1} of SO_2 and 9.4×10^{10} mol-y^{-1} of NO_x transported

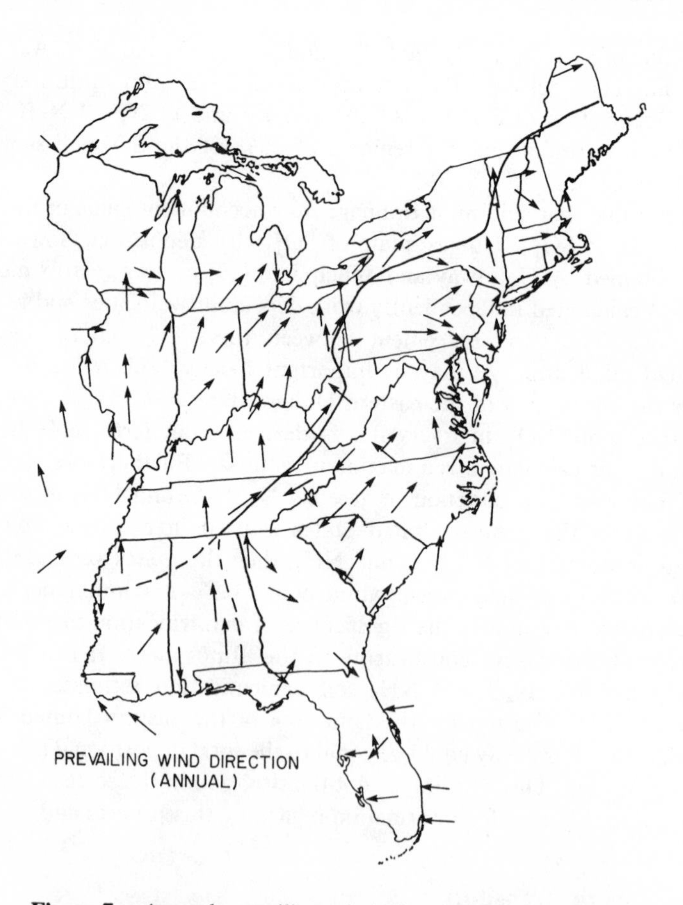

Figure 7. Annual prevailing wind directions in the eastern United States. The heavy line separates areas in which winds blow toward Canada from those in which winds blow toward the Atlantic Ocean or Gulf of Mexico.

to Canada, and 9.4×10^{10} mol-y^{-1} of SO_2 and 11.5×10^{10} mol-y^{-1} of NO_x blown over the Atlantic. Galloway and Whelpdale [19] estimated the total wind blown flux of SO_2 from the eastern United States as 18.5×10^{10} mol-y^{-1}, a value in excellent agreement with our atmospheric transport estimate. However, their estimates for the fluxes to Canada and the Atlantic are 6.3×10^{10} and 12.2×10^{10} mol-y^{-1} of SO_2, respectively. Accounting for the discrepancy between their and our estimates is difficult, because reasonable errors in estimation (the Galloway and Whelpdale estimates are within a factor of two of ours) cover the range of variation between estimates.

DISCUSSION AND CONCLUDING REMARKS

A major purpose of this chapter was to develop an approach, the geochemical mass balance model, to evaluation of the regional eastern U.S. problem of acid precipitation. Fluxes estimated for the model are summarized in Table V and Figure 8. Some caution should be exercised in interpreting the significance of these estimates. Wet deposition and runoff are based on data for a one-year, rather than multiyear, period. Difference calculations leading to estimates of dry deposition are based on estimates for pristine sulfur and nitrogen in river runoff, which could be determined more accurately by careful examination of soil and rock types throughout the eastern United States. Probable time lags between inputs and outputs have not been considered, e.g., storage of acid components in soil before release to rivers. As the data base is expanded, however, and the processes involved in acid precipitation are more fully understood, the model can be refined. Initial

Table V. Estimated Annual Fluxes of Acid Precipitation Constituents S and N for the Eastern United States (Method of Calculation Is Included for Fluxes Estimated by Difference)

Flux	*S* *(10^{10} mol)*	*N* *(10^{10} mol)*
A. Emission	35.6	27.7
B. Wet Deposition	6.7	5.4
C. Present River	40.0	7.8
D. Pre-man River	24.0	1.0
E. Anthropogenic River (C-D)	16.0	6.8
F. Dry Deposition (E-B)	9.3	1.4
G. Atmospheric Transport (A-E)	19.6	20.9
H. Atmospheric Transport to Canada (0.52 × G for S) (0.45 × G for N)	10.2	9.4
I. Atmospheric Transport to the Atlantic (0.48 × G for S) (0.55 × G for N)	9.4	11.5

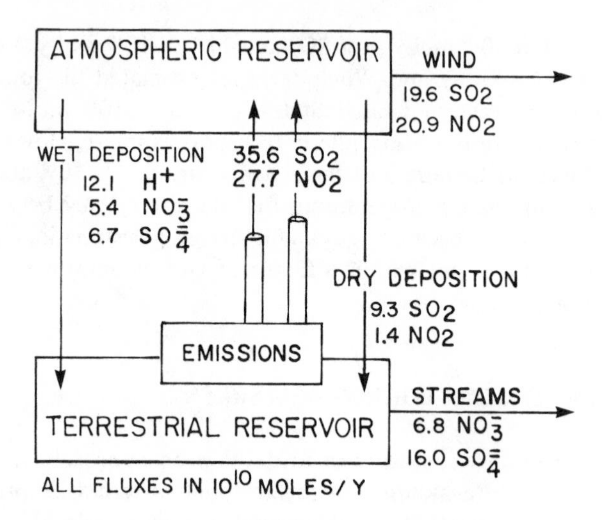

Figure 8. Geochemical mass balance model for constituents of acid precipitation with values for the eastern United States included.

refinement might include the use of individual drainage basins, instead of states, for deriving some of the initial flux estimates. Transient-state considerations also might be introduced into the modeling effort, particularly for the prediction of the future of acid precipitation. Whatever the case, the excellent agreement between the sulfur flux estimates of Galloway and Whelpdale [19] and those obtained from our model attest to the overall validity and value of the geochemical mass balance model.

A number of conclusions can be drawn from the modeling effort. It is apparent that the major inputs of SO_2 and NO_x into the atmosphere of the eastern United States are anthropogenic. Biogenic sources and anthropogenic fluxes from outside the region probably are insignificant. Atmospheric transport is the main removal process, accounting for about 55 and 75% of the SO_2 and NO_x emissions, respectively. To establish the ultimate destination of emissions removed by wind, careful monitoring, tracing and modeling of atmospheric circulation patterns and acid constituent loads are needed. Similar studies are necessary to evaluate the amount of NO_x emissions in the western United States that are transported to the eastern United States.

Other refinements of nitrogen fluxes are needed. The effects of fertilizers and nitrogen constituents of acid precipitation on the release of NO_x from agricultural and natural soils should be quantified experimentally. Also, measurements of rates of denitrification processes leading to the uptake of NO_3^- are necessary to determine if these processes are an important sink for nitrogen constituents of acid precipitation.

Dry and wet deposition respectively account for 26 and 19% of the SO_2 emissions and 5 and 19% of the NO_x emissions. Because these fluxes might cause the most immediate ecosystem responses, continued, coordinated monitoring of wet deposition and implementation of a similar system to monitor dry deposition are necessary. Thorough understanding of the acid precipitation problem in the eastern United States also requires continued monitoring of the dissolved constituent load of major rivers in this region.

Acid precipitation over the eastern United States can react with rock and soil minerals according to reactions like the following:

$$2CaCO_3 + H_2SO_4 = 2Ca^{2+} + 2HCO_3^- + SO_4^{2-}$$

and

$$Mg_5Al_2Si_3O_{10}(OH)_8 + 5H_2SO_4 = Al_2Si_2O_5(OH)_4 + 5Mg^{2+} + 5SO_4^{2-} + H_4SiO_4 + 5H_2O$$

Reactions of this nature involve both "acids" (H_2SO_4 and HNO_3) and a variety of carbonate, aluminous and silicate minerals [29,30]. It should be remembered, however, that in a region as large as the eastern United States, carbonic acid (H_2CO_3) is the principal weathering acid.

The complete neutralization of excess acid in acid precipitation over the eastern United States because of reactions with rock and soil minerals, according to the model, results in a maximum stream discharge of 16×10^{10} mol-y^{-1} of SO_4^{2-} and 6.8×10^{10} mol-y^{-1} of NO_3^-. The amounts of cations released by weathering reactions similar to those above depend on the initial reactant mineral compositions and final degradation products. By writing all likely equations for weathering reactions, we can show that the probable maximum molar ratios of cations produced to acid (H^+) consumed is 2:1 for SO_4^{2-} and 1:1 for NO_3^-. Thus, model calculations show that about 40×10^{10} μeq/yr of cations could be released by weathering of soil and rock minerals by the components in acid precipitation. This value corresponds to a cation denudation rate of 180 meq-m^{-2}-y^{-1} for the eastern United States, 15% of the overall rate of 1200 meq-m^{-2}-y^{-1}. The latter value is based on average cation concentrations for 120 rivers in the eastern United States of 622 μM Ca^{2+}, 279 μM Mg^{2+}, 453 μM Na^+ and 50 μM K^+, a total river discharge of 1.16×10^{15} L and an area of 2.22×10^{12} m^2.

The usefulness of a regional mass balance model in describing and evaluating fluxes of acid precipitation constituents has been demonstrated. Models of this nature, although perhaps not unique solutions to a problem, can be useful in organizing and compiling information, identifying processes, and quantifying source and sink fluxes. This type of model can be used to evaluate impacts of man-related chemical substances (e.g., fertilizers and pesticides) other than acid precipitation, on a global and regional scale.

ACKNOWLEDGMENTS

Primary funding for this research was provided by the Office of Research and Development, U.S. Environmental Protection Agency, under Cooperative Agreement No. CR 807856 01. Additional funding was provided by Cornell University. This publication is ERC Report No. 18. The Ecosystems Research Center (ERC) was established in 1980 as a unit of the Center for Environmental Research at Cornell University. The work and conclusions published herein represent the views of the authors, and do not necessarily represent the opinions, policies or recommendations of the Environmental Protection Agency or of Cornell University. The authors gratefully acknowledge Laurie Steigely and Linda Bischoff for typing the manuscript and Jane Schoonmaker for reviewing it. Hawaii Institute of Geophysics contribution No. 1369.

REFERENCES

1. Garrels, R.M., F.T. Mackenzie and C. Hunt. *Chemical Cycles and the Global Environment.* (Los Altos, CA: W. Kaufman, Inc., 1975).
2. Lerman, A., F.T. Mackenzie and R.M. Garrels. "Modeling of Geochemical Cycles: Phosphorus as an Example," *Geol. Soc. Am. Mem.* 142 (1975), pp. 205–218.
3. Svensson, B.H., and R. Söderlund, Eds. *Nitrogen, Phosphorus and Sulfur—Global Cycles, Ecol. Bull.* Vol. 22 (1976).
4. Bolin, B., E.T. Degens, P. Duvigneaud and S. Kempe. "The Global Biogeochemical Carbon Cycle," in *The Global Carbon Cycle,* B. Bolin, E.T. Degens, S. Kempe and P. Ketner, Eds., SCOPE Rept. 13 (New York: John Wiley & Sons, Inc., 1979), pp. 1–56.
5. "1977 National Emissions Report," EPA Report 450/4–80–005, National Emissions Data Systems, Office of Air and Waste Management, U.S. EPA (1980).
6. "1976 National Emissions Report," EPA Report 450/4–79–019, National Emissions Data Systems, Office of Air and Waste Management, U.S. EPA (1979).
7. "1975 National Emissions Report," EPA Report 450/2–78–020, National Emissions Data Systems, Office of Air and Waste Management, U.S. EPA (1978).
8. "1974 National Emissions Report," EPA Report 450/2–78–026, National Emissions Data Systems, Office of Air and Waste Management, U.S. EPA (1978).
9. "1973 National Emissions Report," EPA Report 450/2–76–007, National Emissions Data Systems, Office of Air and Waste Management, U.S. EPA (1976).
10. "1972 National Emissions Report," EPA Report 450/2–74–012, National Emissions Data Systems, Office of Air and Waste Management, U.S. EPA (1974).
11. "Climatic Atlas of the United States," U.S. Dept. of Commerce, Env. Sci. Serv. Adm., Env. Data Ser. (1968).
12. "National Air Quality and Emissions Trends Report 1976," EPA Report 450/1–77–002, Office of Air Quality, U.S. EPA (1977).
13. Likens, G.E., R.F. Wright, J.N. Galloway and T.J. Butler. "Acid Rain," *Sci. Am.* 241:43–51 (1979).
14. "NADP Data Report, Precipitation Chemistry, Vol. 2, Part 4," Natural Resource Ecology Laboratory, Colorado State University, Ft. Collins, CO (1979).
15. "NADP Data Report, Precipitation Chemistry, Vol. 3, Parts 1–3," Natural Resource Ecology Laboratory, Colorado State University, Ft. Collins, CO (1980).
16. Meybeck, M. "Pathways of Major Elements from Land to Ocean through Rivers," in *River Inputs to Ocean Systems,* J.M. Martin, J.D. Burton and D. Eisma, Eds. (Switzerland: UNEP and UNESCO, 1981), pp. 18–30.

17. Wollast, R. "Interactions in Estuaries and Coastal Waters," in *Biogeochemical Cycles,* G. Likens, Ed., SCOPE Rept. 16 (New York: John Wiley & Sons, Inc., 1981).
18. Clarke, F.W. "The Composition of the River and Lake Waters of the United States," *USGS Professional Paper* 135 (1924).
19. Galloway, J.N., and D.M. Whelpdale. "An Atmospheric Sulfur Budget for Eastern North America," *Atmos. Environ.* 14:409–417 (1980).
20. Robinson, E., and R.C. Robbins. "Gaseous Nitrogen Compound Pollutants from Urban and Natural Sources," *J. Atmos. Poll. Control. Assoc.* 20:303–306 (1970).
21. Söderlund, R., and B.H. Svensson. "The Global Nitrogen Cycle," *Ecol. Bull.* 22:23–73 (1976).
22. Burns, R.C., and R.W. Hardy. *Nitrogen Fixation in Bacteria and Higher Plants* (New York: Springer-Verlag, 1976).
23. Garrels, R.M., F.T. Mackenzie and C. Hunt. *Chemical Cycles and the Global Environment,* 2nd ed. (Los Altos, CA: W. Kaufman, Inc., in press).
24. Chamberlain, J., H. Foley, D. Hammer, G. MacDonald, D. Rothaus and M. Ruderman, Eds. *The Physics and Chemistry of Acid Precipitation,* Tech. Rept. JSR-81-25 (Arlington, VA: SRI International, 1981).
25. Makarov, B.N. "Liberation of Nitrogen Dioxide from Soil," *Pochvodenie* 1:49–53 (1969).
26. Kim, C.M. "Influence of Vegetation Type on the Intensity of Ammonia and Nitrogen Dioxide Liberation from Soil," *Soil Biol. Biochem.* 5:163–166 (1973).
27. Galloway, J.N., and G.E. Likens. "Acid Precipitation: The Importance of Nitric Acid," *Atmos. Environ.* 15:1081–1085 (1981).
28. Zuehls, E.E., G.L. Ryan, D.B. Peart and K.K. Fitzgerald. "Hydrology of Area 35, Eastern Region, Interior Coal Province, Illinois and Kentucky," USGS Open File Rept. 81–403 (1981), pp. 36–37.
29. Johnson, N.M., C.T. Driscoll, J.S. Eaton, G.E. Likens and W.H. McDowell. " 'Acid Rain', Dissolved Aluminum and Chemical Weathering at the Hubbard Brook Experimental Forest, New Hampshire," *Geochim. Cosmochim. Acta* 45(9):1421–1427 (1981).
30. Glass, N.R., D.E. Arnold, J.N. Galloway, G.R. Hendry, J.J. Lee, W.W. McFee, S.A. Norton, C.F. Powers, D.L. Rambo and C.L. Schofield. "Effects of Acid Precipitation," *Environ. Sci. Technol.* 16(3):162–169 (1982).

CHAPTER 2

Chronology, Magnitude and Paleolimnological Record of Changing Metal Fluxes Related to Atmospheric Deposition of Acids and Metals in New England

J.S. Kahl
S.A. Norton
J.S. Williams

The chemistry of nondystrophic oligotrophic lakewaters in New England is controlled primarily by bedrock lithology and atmospheric deposition. Secondarily, soil properties, hydrology and in-lake biological processes are often important. Surface water pH values covary with Na, K, Ca and Mg concentrations, which are determined by the availability of easily weathered minerals in the soil or bedrock, and by groundwater residence time in chemically resistant terranes. The resultant pH affects the solubility of Fe, Mn and Al hydroxides, and Zn and Cd adsorption/desorption reactions, and therefore determines the concentrations of these and chemically similar elements in surface waters [1].

No long-term studies of the chemistry of lakewaters in New England are available to evaluate long-term effects of atmospheric deposition of increasing amounts of acids and metals on lakewater chemistry. A lake sediment record, in conjunction with an understanding of sediment–water interactions, enables us to reconstruct the chronology of increasing atmospheric deposition of acids and metals, in lieu of a long-term precipitation and surface water chemical record. Reported here is an interpretation of sediment chemistry profiles from three oligotrophic, clearwater, acidic lakes with undisturbed drainage basins. In addition, initial findings of experimental studies designed to determine the influence of recent lake acidification on these sediment profiles are presented.

Acidification of precipitation (and subsequently groundwater, streams and lakes) with a strong acid (H_2SO_4 and/or HNO_3) should result in increased release of all cations from desorption reactions or weathering. For example, Hanson et al. [2] found that in areas receiving lower-pH precipitation in the area from New England to the Gaspé Peninsula, Quebec, the organic soils of spruce-fir forests had significantly lower concentrations of Ca, Mg, Zn and Mn. This depletion was attributed to accelerated leaching of these metals.

Superimposed on accelerated chemical weathering responses to precipitation acidification are the effects of increased atmospheric loading of metals (especially Pb and Zn). Numerous studies [3–6] of lake sediment cores infer that atmospheric deposition of Pb, Zn and other trace metals has recently increased significantly in eastern North America. Chemical analysis of dated sediment shows that the trace metal flux to sediments increased at least 100 years ago [5,7]. The atmospheric deposition of metals varies regionally; locally, metal enrichment in the sediment is a function of sedimentation rate, background metal concentration, sediment focusing and other often poorly understood factors.

Elevated levels of Pb are found in soils receiving deposition of acids and heavy metals [2,8,9]. Although archived soils are not available for comparative studies, Siccama and Smith [9] report increases in Pb over a 16-year period in forest soils in Massachusetts. Hanson et al. [2] report a gradient of Pb content in the organic matter of subalpine fir forest organic soils from southern Vermont (high Pb) to the Gaspé Peninsula (low Pb). The precipitation pH gradient ranges from about 4.1 (weighted annual mean) in southern Vermont [10] to about 4.6 in the Gaspé area [11]. Data from Semonin et al. [10] in conjunction with Lazrus et al. [12] suggest that trace metal deposition correlates with acid deposition.

Sediment deposited in the deeper part of lakes is derived in part from the terrestrial ecosystem and resuspended littoral sediments. Thus, as precipitation becomes more acidic and metal-laden, the sediments should reflect terrestrial changes—decreased concentrations of easily mobilized metals (Zn, Mn, Ca and Mg)—and increased concentration of Pb. Acidification of surface water may also cause chemical changes in this detritus before sedimentation, as well as increased in situ leaching of lake and stream sediments.

METHODS

Lake Selection

Little Long Pond, Dream Lake and Unnamed Pond (Figure 1) (summer pH 5.5, 4.5, and 4.5, respectively) are typical acidified lakes from a larger group of New England and Scandinavian lakes that have been studied in the past six years. However, their respective topographies and bathymetries (Table I) result in somewhat different sediment chemistry. The criteria for lake selection were:

1. Noncalcareous bedrock: the three ponds are situated in till or glacial outwash, on granite bedrock.

2. Relatively undisturbed watershed: none of the watersheds has been lumbered in the past 50–100 years, and none has any human habitations in the drainage basin.
3. No major geographic or anthropogenic dissimilarities, such as proximity to industrial plants or to saltwater (localized atmospheric input).

Paleolimnology

Sediment cores (0.4–0.6 m) were obtained from the deepest point of each lake with a 10-cm-diameter piston corer [13] operated from a raft. All cores that were used had undisturbed sediment/water interfaces and intact stratigraphies. The cores were sectioned in the field in 0.5-cm increments from 0 to 20 cm depth in the core and in 1.0-cm increments below 20 cm depth. The samples were stored in plastic bags in the dark at 4° C until processing.

Sediment for metal analysis is dried at 110° C for 24 h, crushed, and ashed

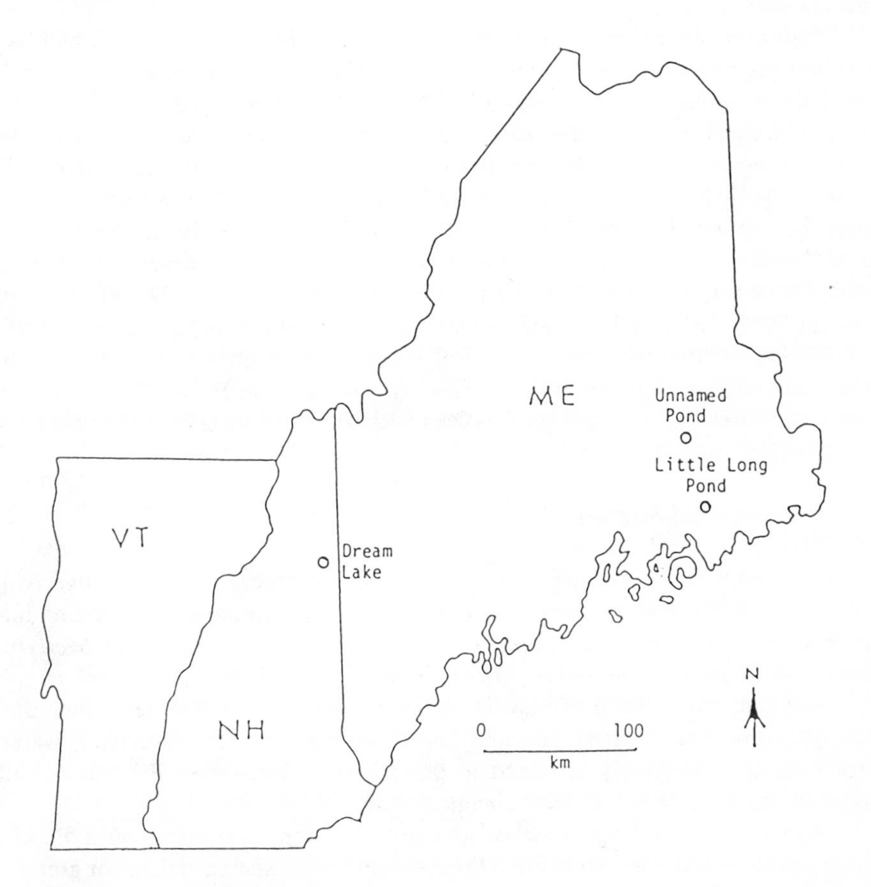

Figure 1. Locations of study lakes.

Table I. Lake Characteristics

	Drainage Basin Area (ha)	*Lake Area (ha)*	*Maximum Depth (m)*	*Altitude (m)*	*Longitude*	*Latitude*
Dream Lake, NH	25	3.7	1.5	792	71°06′ E	44°27′ N
Little Long Pond, ME	145	31	24	72	68°05′ E	44°38′ N
Unnamed Pond, ME	8 (Kettle)	8	7	141	68°11′ E	45°12′ N

at 550° C for 3 h. A 100-mg aliquot of ashed sediment is acid-digested in a HF–aqua regia solution for 1 h at 90° C, and brought to 100 mL volume in 6% v/v HF, 1% v/v aqua regia and 5.6% w/v H_3BO_3 (to dissolve insoluble metal fluoride salts) [14].

Sediment chronologies were determined using the nuclide ^{210}Po, which is a daughter product in the decay series involving ^{222}Rn. ^{222}Rn is released from rocks and soils as a result of the decay of ^{226}Ra. ^{210}Pb is scavenged by precipitation (maximum residence time one week) [15] and incorporated rapidly into soils [16] or onto suspended particulates in aqueous media [17]. Terrestrial lead is chemically immobile [18]; most Pb deposited in sediments has been demonstrated to have fallen directly on the lake [17]. A steady-state delivery of ^{210}Pb to the sediment should produce an exponential decay (concentration) curve, which decreases to nearly background in approximately 100 y (the half life for ^{210}Pb is 22.2 y). Sediment used for ^{210}Pb dating is first heated to 600° C to extract the granddaughter (^{210}Po) of ^{210}Pb [19]. ^{210}Po (and reference ^{208}Po) is collected on glass wool, leached with nitric acid and put into an HCl solution by heating. The Po is then autoplated onto silver discs and counted for 20,000 s with an Ortec surface barrier detector.

Experimental Studies

Sediment/water cores (6.35 cm diameter; 15 cm of sediment with 1 L of overlying water) were obtained with scuba for pH manipulation studies to assess metal mobilization from sediments. These cores were aerated at 4–6° C; the pH of overlying water was adjusted with mixed H_2SO_4 + HNO_3 (1 : 1), to achieve several pH values ranging from 3.9 to 6.5 (realistic natural pH values). All treatments, including controls, were done in triplicate and lasted for nine months. Overlying waters were sampled periodically, acidified to pH ≤ 2 in acid-washed LPE bottles, and stored in the dark at 3–5 C until chemical analysis.

Analyses of waters and sediment digestions were performed on a Perkin-Elmer model 703 Atomic Absorption Spectrometer using standard flame or graphite furnace (HGA-2200) techniques. The following parameters were monitored: pH, Ca, Mg, K, Na, Mn, Zn, Pb and Al.

THE SEDIMENT RECORD

Figures 2, 3 and 4 show chemical profiles of profundal sediments of three lakes in northern New England (Figure 1, Table I). Major element concentrations and deposition rates indicate steady-state conditions in Dream and Little Long, or long-term natural changes in Unnamed, where increasing diatom abundance (flux) may have altered relative concentrations of other constituents by dilution with silica. Increases in concentrations of Pb and Zn in these sediments are concurrent with the beginning of decreasing concentrations of Ca and Mn. Based on ^{210}Pb chronology, the increase in heavy metals begins about 1860–1880 (Figure 5), which are reasonable dates for the onset of major industrialization and therefore atmospheric emissions in the United States. Apparent Pb atmospheric inputs to the lakes have continued to increase nearly to the present, giving rise to deposition rates as high as 3.25 $\mu g\text{-}cm^{-2}\text{-}y^{-1}$ (in Little Long, Table II). These sediments

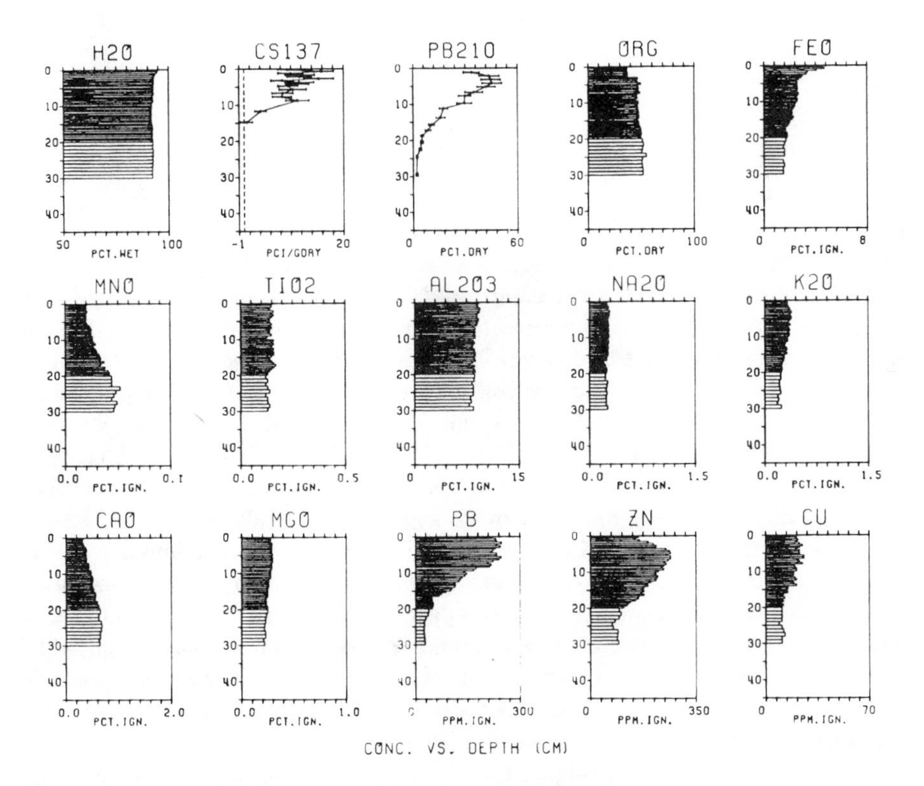

Figure 2. Sediment chemistry: Little Long Pond, Maine.

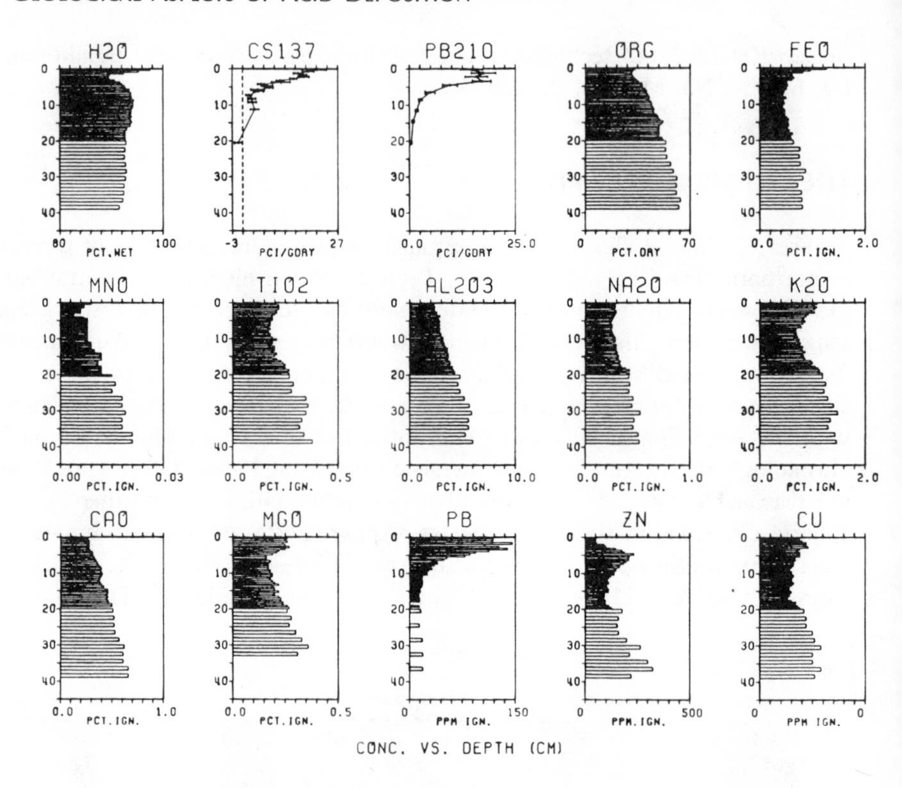

Figure 3. Sediment chemistry: Unnamed Pond, Maine.

have been enriched in Pb by as much as a factor of 10 over background concentrations. Zinc enrichment is as great as threefold (Table II). It is possible that hydrogen ion inputs began increasing concurrently with the heavy metals in the mid-1800s. If this is so, the Ca and Mn decrease in progressively younger sediments (Figure 5) may be due to acid leaching of soil detritus. Zinc shows decreasing sediment concentrations beginning 20–50 years ago, indicating that Zn mobilization from soils may occur at a lower pH than does Ca or Mn mobilization, or that some Zn is leached from recent sediments as the lakes become acidified. Experiments indicate that zinc is more mobile than Pb and may not be deposited in sediments after mobilization if subjected to low ambient pH in the water column.

Maximum sediment deposition rates for Pb and Zn are very similar for individual lakes (Table II), although Zn is generally more concentrated than Pb in precipitation [12]. This discrepancy suggests higher Zn loss from the watershed, which is borne out by surface water chemistry for these lakes; total dissolved Pb is normally ≤ 1 ppb whereas total Zn is normally 5–20 ppb. The total anthropogenic Pb and Zn in Little Long sediment is markedly higher than in the other two ponds (Table II). There is no evidence for higher-than-average atmospheric inputs to the pond,

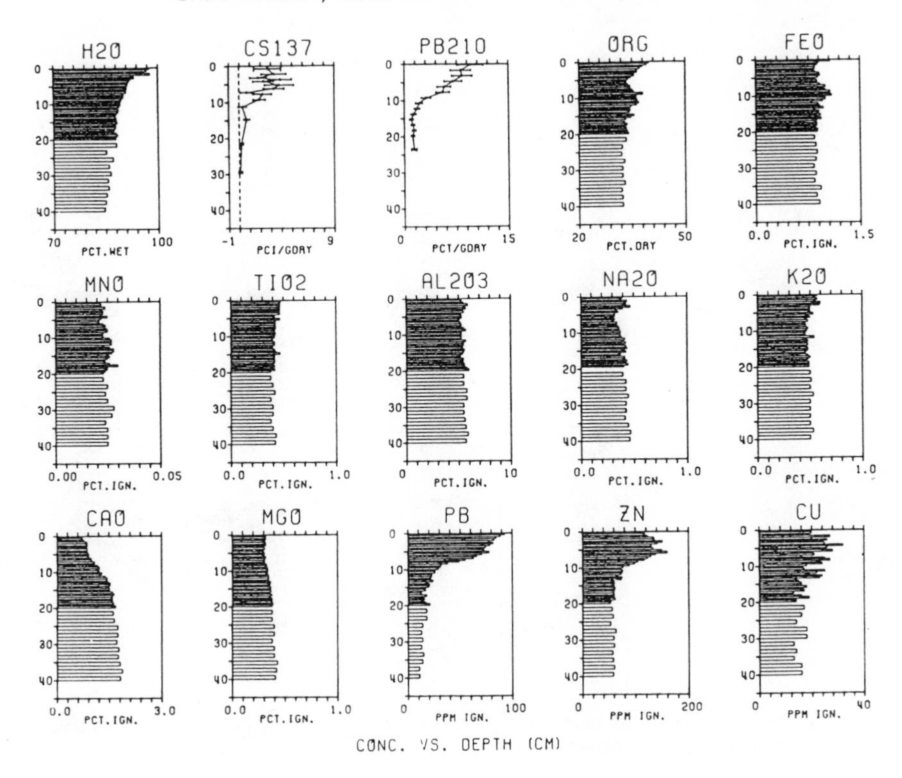

Figure 4. Sediment chemistry: Dream Lake, New Hampshire.

implying that focusing of sediment rich in Pb and Zn occurs in Little Long, resulting in higher Pb and Zn concentrations in deep water sediments. Consistent with this, littoral sediments are impoverished in Pb relative to profundal sediments (Figure 6). It is likely that the total anthropogenic Pb input per square meter for the entire lake bottom is nearly the same for all three ponds (as suggested by Dillon and Evans [20] for some Canadian lakes) but that it is distributed differently within each lake. Dillon and Evans [20] also report that increasing Pb concentrations are correlated with increasing depth in lakes. However, redistribution of bulk sediment does not alter concentration profiles. There must be a chemical focusing or a particulate fractionation for Pb (there may be different processes for other metals) that result in Pb-rich sediments being deposited in bathymetric lows. For example, atmospherically deposited Pb may be attached to, or rapidly scavenged by, small particulates in the water, which are preferentially deposited in deeper water. Of the elements investigated, only Pb is strongly focused.

The variability of sediment chemical profiles from different depths within the same lake (e.g., this study and Dillon and Evans [20]) suggests that great caution must be taken when generalizing from one core to an entire lake.

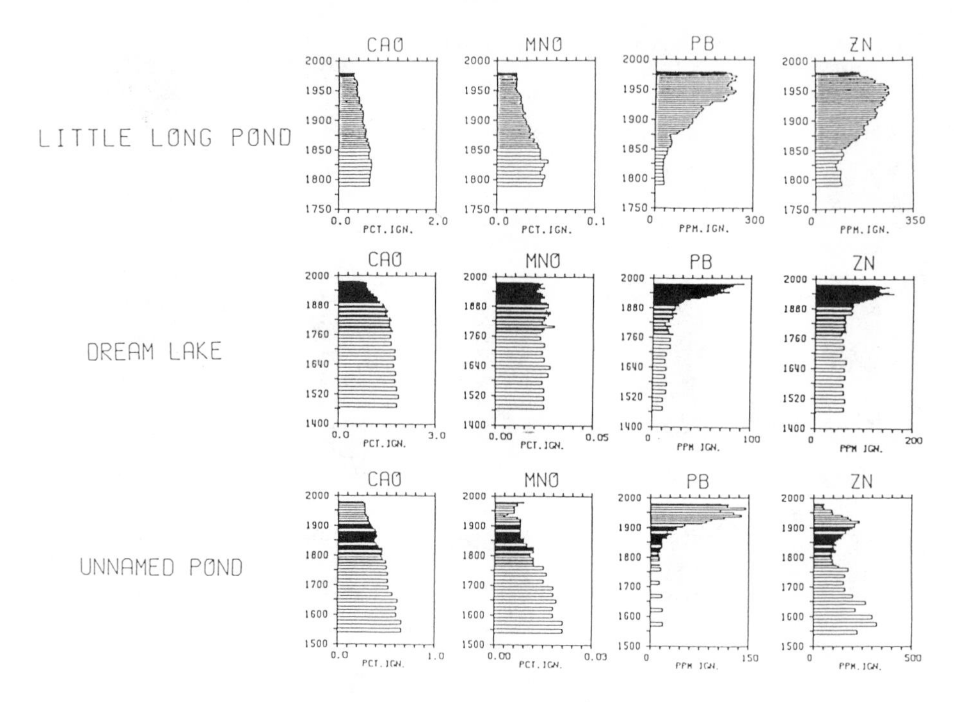

Figure 5. Selected sediment chemistry profiles vs time in study lakes. Dates are based on ^{210}Pb chronology as discussed in text.

DISCUSSION

Terrestrial Leaching

The decline in sediment concentration of Ca and Mn in the three lakes is generally concurrent with the increased loading of Pb and Zn, which is ubiquitous and nearly synchronous (1860–1880) across northern New England. Whether loss of these elements from the detritus occurs on the terrestrial landscape or in contact with stream or lakewater is not well understood.

In essentially all clearwater lakes in the northeastern United States, the water has a significantly higher pH than does the precipitation, due to chemical weathering reactions between soils or bedrock and groundwater. Experimental evidence in the field [21] and in the laboratory [22] suggests that acidification of precipitation increases the rate of leaching of Ca, Mg, Mn and other elements from soils. Temporal and spatial studies of percent base saturation in soils also suggest that acidification of precipitation results in a loss of exchangeable cations from soils [23], consistent with the experimental studies. If one assumes that the data for a sediment core are typical for the whole lake and its drainage basin (a risky assumption) and if the "lost" metals in the sediment core are put into solution through time, whether from terrestrial or subaqueous leaching, the estimated increase in Ca con-

Table II. Pb and Zn Inputs[a]

	Total Anthropogenic Inputs (μg-cm^{-2})		*Maximum Net Deposition Rate* (μg-cm^{-2}-y^{-1})		*Maximum Sediment Enrichment*	
	Pb	*Zn*	*Pb*	*Zn*	*Pb*	*Zn*
Dream Lake	45	57	0.99	1.23	7X	3X
Little Long Pond	197	195	3.25	2.74	10X	3X
Unnamed Pond	54	48	1.17	1.28	8X	2X

[a] Total anthropogenic metal is calculated from sediment concentration profiles (water content, organic content and metal concentration) as the total micrograms of metal in excess of background in a 1- × 1-cm square column of sediment. Maximum net deposition is the greatest observed quantity above background for a particular yearly "slice" of sediment. Maximum sediment enrichment is the greatest ratio obtained (for any year) of total grams of metal divided by background grams of metal.

centration in surface waters would be on the order of 0.1 ppm or less. Other elements would have smaller absolute increases but possibly significant relative increases (e.g., the missing Mn from Little Long sediment profiles would yield an increase of approximately 20 ppb). The loss of Ca and Mn from detritus, as indicated by core studies, started in some lake/drainage basins as early as the first elevated concentrations of Pb and Zn. It seems plausible that accelerated terrestrial leaching of detritus would precede lake acidification, and that subaqueous leaching would not occur until lake acidification occurs.

The profile of Zn in recent sediments is a result of two competing processes: (1) increased atmospheric loading with time and (2) greater terrestrial and aquatic leaching due to increasingly acidic precipitation. Other lakes in northern New England (not reported here) with higher pH (>6.0) typically have Zn sediment

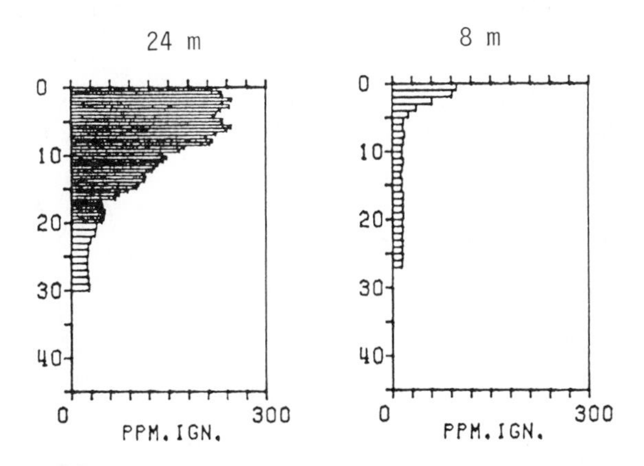

Figure 6. Concentration profiles of lead from 24- and 8-m depths in Little Long Pond, Maine.

concentrations that increase to the present. Because the terrestrial detritus in New England is subjected to precipitation with weighted mean pH ≤ 4.4 [10], the elevated concentrations of Zn in recent sediments of higher pH lakes may be due to:

1. direct atmospheric deposition of Zn directly on the lake and subsequent sedimentation (in contrast to Zn behavior in lower pH lakes);
2. back reaction of detritus (adsorption of Zn?) with the higher pH water; or
3. a lack of leaching of sediment by lakewater.

In acidified lakes, the recent decreases may be augmented by sediment leaching or by decreased sediment scavenging.

pH Manipulation Studies

We have manipulated the pH of the overlying water of 90 sediment/water cores taken from five lakes, including Little Long and Unnamed, to investigate possible effects of acidification on sediment and water chemistries. Our sediment chronology in conjunction with sediment chemistry presented here argues that precipitation has become progressively more "acidic" during the last 100 years, and may have been leaching soil detritus (with subsequent effects in sediments) since the mid-1800s. More recently, lakes have become acidified and the possibility of in-lake leaching has increased. Our experimental studies were to determine if in situ chemical changes may be partially responsible for the sediment profiles presented here.

Typical data from these microcosms are presented in Table III. The order of percentage increase in overlying waters is (for treatment pH 3.9) Al > Mn > Zn > Pb > Ca > Mg > K > Na. The total amount of metal released from these sediments, expressed as a percent of the total amount of the metal in the top 2 cm of the core, may be sufficient to appreciably alter sediment chemistry for some metals (Table IV). The order of release from sediments expressed as a

Table III. Changes in Concentration of Dissolved Metals in Overlying Water of pH-Controlled Sediment/Water Cores (8 m depth) from Little Long Pond, Maine

Initial	*pH 3.9 9 Months*	*Percent Increase*	*pH 5.0 9 Months*	*Percent Increase*
Al (ppb) 50	2900	5700	105	110
Zn (ppb) 20	195	875	34	70
Mn (ppb) 30	1170	3800	10	−67
Pb (ppb) 0.5	3.9	680	0.5	0
Ca (ppm) 1.5	9.6	540	3.1	106
Mg (ppm) 0.4	1.3	215	0.7	73
K (ppm) 0.4	1.1	165	0.7	73
Na (ppm) 2.0	3.7	85	3.4	70

Table IV. Percent of Metal Reservoir Released from Top 2 cm of Sediment of Experimental Cores from Little Long Pond, Maine after 9 Months

	Al	*Ca*	*K*	*Mg*	*Mn*	*Na*	*Pb*	*Zn*
Littoral Cores (8 m)								
pH 3.9	0.8	22.5	1.4	5.2	33.7	1.4	0.4	14.2
pH 5.0	<0.1	3.2	0.2	1.4	<0.1	0.2	<0.1	4.0
Profundal Cores (24 m)								
pH 3.9	2.4	51.3	4.7	8.6	50.7	4.5	0.5	21.1
pH 5.0	0.1	2.6	1.2	1.7	<0.1	0.2	<0.1	4.0

percentage of the sediment pool is Ca > Zn > Na ? Mg > K > Mn > Al > Pb for pH <5.0, although Mn release increases dramatically as the pH approaches 4.0. Aluminum profiles are unlikely to be perceptibly altered, due to the very large Al pool, but smaller absolute releases of Zn or Mn may be quantitatively important for trace metal profiles.

We have pointed out that concentrations of Ca, Mn and Zn decrease in younger sediments of acidic lakes. However, Ca and Mn have decreased in progressively younger sediments for at least 100 years. Assuming that lake acidification is a relatively recent event, these profiles have probably not been greatly influenced by pH-dependent diagenetic processes. Ca and Mn sediment profiles may reflect a steady decline in soil base saturation (and thus the decline in the acid-neutralizing capacity of the soil). Soils in these watersheds eventually have become sufficiently leached in conjunction with increasingly acidic precipitation to no longer function effectively as acid neutralizers, and lake acidification has ensued. We suggest that lakewaters with pH 5.0–5.5 leach Zn from sediments, and/or prevent further accumulation. Experimental work indicates that substantial release of metals from sediments at realistic pH values is possible and thus sediments may serve as a source of dissolved metals and act as acid neutralizers.

SUMMARY

Zinc and lead concentrations and deposition rates in sediments of three New England lakes began increasing (but at different rates) ca. 1860–1880. Maximum concentration increases range from 2 to 10 times background. Sediment concentrations of Ca and Mn decline concurrently with the increase in heavy metals. These processes are attributed to increased loading of metals and acids from precipitation since that time. Maximum deposition rates of Pb and Zn range from 1 to 3 $\mu g\text{-}cm^{-2}\text{-}y^{-1}$ for the three lakes. Maximum deposition rates for Pb occur in the youngest sediment. However, recent sediments are depleted of Zn in spite of increased atmospheric loading. Watershed contribution and in-lake focusing processes cause non-uniform deposition of metals. Allochthonous alkali-leached but heavy-metal-laden detritus has strongly influenced sediment chemistry in acidified watersheds for

$\geq$ 100 years. Additionally, lakewater–sediment interactions may sensibly alter sediment chemistry (Ca, Mn, and possibly Zn) and water chemistry (Al, Mn, and Zn) as the lakes become acidified.

ACKNOWLEDGMENTS

Our research has been supported by the National Science Foundation (Grant DEB-7810641 to R.B. Davis and S.A. Norton), the Office of Water Research and Technology of the U.S. Department of the Interior (Grant A-053 to J.S. Kahl and S.A. Norton), the Fish and Wildlife Service of the U.S. Department of the Interior (Contract #14-16-0009-79-040 to S.A. Norton and R.B. Davis), the U.S. Environmental Protection Agency (Contract #APP-0054-1980 to S.A. Norton and R.B. Davis) and the Graduate Student Board at the University of Maine (grant to J.S. Kahl). Tom Robertson and numerous undergraduate students assisted in the field work. Marilyn Morrison and Elaine Hall gave great assistance in field work and in the chemical analyses. Geneva Blake determined the ^{210}Pb chronology, and Denis Hanson developed the computer programs that made many of our interpretations possible. Two anonymous reviewers made many helpful suggestions to improve the manuscript.

REFERENCES

1. Norton, S.A., R.B. Davis and D.F. Brakke. "Responses of Northern New England Lakes to Atmospheric Inputs of Acids and Heavy Metals," Completion report A-048-ME, Land and Water Resources Center, University of Maine (1981).
2. Hanson, D.W., S.A. Norton and J.S. Williams. "Modern and Paleolimnological Evidence for Accelerated Leaching and Metal Accumulation in Soils in New England, Caused by Atmospheric Deposition," *Water, Air Soil Poll.* (1982).
3. Norton, S.A., C.T. Hess and R.B. Davis. "Rates of Accumulation of Heavy Metals in Pre- and Post-European Sediments in New England Lakes," in *Atmospheric Pollutants in Natural Waters,* S.J. Eisenreich, Ed. (Ann Arbor, MI: Ann Arbor Science Publishers, Inc., 1981), pp. 409–421.
4. Galloway, J.N., and G.E. Likens. "Atmospheric Enhancement of Metal Deposition in Adirondack Lake Sediments," *Limnol. Oceanog.* 24(3):427–433 (1979).
5. Evans, R.D., and P.J. Dillon. "Historical Changes in Anthropogenic Lead Fallout in Southern Ontario," in *Proceedings of the 2nd Symposium on Interactions Between Sediment and Freshwater* (Stuttgart: E. Schweizerbart'sche Verlagsbuchhandlung, 1982).
6. Davis, R.B., S.A. Norton, C.T. Hess and D.F. Brakke. "Paleolimnological Reconstruction of the Effects of Atmospheric Deposition of Acids and Heavy Metals on the Chemistry and Biology of Lakes in New England and Norway," in *Proceedings III Intern-Symposium Paleolimnology Devel. Hydrobiol.,* J. Merilainen, R.W. Battarbee and P. Huttunen, Eds. (The Hague: E. Junk, 1982).
7. Johnston, S.E., S.A. Norton, C.T. Hess, R.B. Davis and R.S. Anderson. "Chronology of Atmospheric Deposition of Acids and Metals in New England, Based on the Record

in Lake Sediments," paper presented at the *Symposium on Acidic Precipitation* sponsored by the Environment Chemistry Division, American Chemistry Society, March 28–April 2, 1982.

8. Andresen, A.M., A.H. Johnson and T.G. Siccama. "Levels of Lead, Copper, and Zinc in the Forest Floor in the Northeastern United States," *J. Environ. Qual.* 9(2):293–296 (1980).
9. Siccama, T.G., and W.H. Smith. "Changes in Lead, Zinc, Copper, Dry Weight, and Organic Matter Content of the Forest Floor of White Pine Stands in Central Massachusetts over 16 Years," *Environ. Sci. Technol.* 14:54–56 (1980).
10. Semonin, R.G., V.C. Bowersox, D.F. Gatz, M.E. Peden and G.J. Stensland. "Study of Atmospheric Pollution Scavenging," SWS Contract Report 252 Illinois Institute of Natural Resources, Urbana, IL (1981).
11. Berry, R.L. "An Assessment of the CANSAP Project after Two Years of Operation," LRTAP79-9 Atmospheric Environment Service, Downsview, Ontario (1979).
12. Lazrus, A.L., E. Lorange and J.P. Lodge, Jr. "Lead and Other Metal Ions in United States Precipitation," *Environ. Sci. Technol.* 4(1):55–58 (1970).
13. Davis, R.B., and R.W. Doyle. "A Piston Corer for Upper Sediment in Lakes," *Limnol. Oceanog.* 14:643–648 (1969).
14. Buckley, D.E., and R.E. Cranston. "Atomic Absorption Analyses of 18 Elements from a Single Decomposition of Alumino-silicate," *Chem. Geol.* 7:273–284 (1971).
15. Poet, S.E. "Lead-210, Bismuth-210, and Polonium-210 in the Atmosphere: Accurate Ratio Measurement and Application to Aerosol Residence Time Determination," *J. Geophys. Res.* 77:6515–6527 (1972).
16. Benninger, L.K., D.M. Lewis and K.K. Turekian. "The Use of Natural Pb-210 as a Heavy Metal Tracer in the River-Estuarine System," in *Marine Chemistry in the Coastal Environment,* ACS Symposium Series No. 18, pp. 202–210 (1975).
17. Farmer, J.G. "The Determination of Sedimentation Rates in Lake Ontario Using the Pb-210 Dating Method," *Can. J. Earth Sci.* 15:431–437 (1978).
18. Andersson, A. "Heavy Metals in Swedish Soils: On Their Retention, Distribution, and Amounts," *Swedish J. Agric. Res.* 7:7–20 (1977).
19. Eakins, J.D., and R.T. Morrison. "A New Procedure for the Determination of Pb-210 in Lake and Marine Sediments, *Int. Appl. Rad. Iso.* 29:531–536 (1978).
20. Dillon, P.J., and R.D. Evans. "Whole-Lake Lead Burdens in Sediments of Lakes in Southern Ontario, Canada," in *Proceedings of the 2nd Symposium on Interactions Between Sediment and Freshwater* (Stuttgart: E. Schweizerbart'sche Verlagsbuchhandlung, 1982).
21. Abrahamsen, G. "Effects of Acid Precipitation on Soil and Forest. 4. Leaching of Plant Nutrients," in *Ecological Impact of Acid Precipitation,* D. Drabløs and A. Tollan, Eds. (Oslo, Norway: SNSF, 1980), p. 196.
22. Cronan, C.S. "Effects of Acid Precipitation on Decomposition and Weathering Processes in Terrestrial Ecosystems," U.S. EPA Acidic Deposition Effects Program Review, Raleigh, NC (1982).
23. Oden, S. "The Acidity Problem—An Outline of Concepts," in *Proceedings of the First International Symposium on Acid Precipitation and the Forest Ecosystems,* L.S. Dochinger and T.A. Seliga, Eds. Gen. Tech. Rept. NE-23 (Washington, DC: U.S. Dept. Agric. For. Serv.) pp. 1–36.

CHAPTER 3

Acid Rain Neutralization by Geologic Materials

Noye M. Johnson

The emergence of "acid rain" as a pervasive, regional phenomenon has led inevitably to the consideration of its consequences. Chief among these concerns has been the chemical effect that strong acids, e.g., sulfur oxides (SO_x) and nitrogen oxides (NO_x), can exert on a landscape. This discussion will address one limited aspect of this chemical question: how is acid rain actually neutralized by a given terrane, and what geochemical principles can be applied to understand and predict the neutralization process?

It was recognized early on that, in areas affected by acid rain, major rivers were not acidified to any appreciable extent, but smaller headwater streams were acidified [1]. It was apparent that between the time (and place) when acid rain falls on the landscape and when (and where) it appears in a major river, it has been neutralized to geologically normal pH levels. This observation was not too surprising, however, because rivers do not course over chemically inert substrates, but rather come into contact with rocks, soils and biologic materials, which have various susceptibility to acid attack. If the landscape were in fact composed exclusively of chemically inert materials, e.g., stainless steel, then river water and acid rain would necessarily be the same chemical solutions. The question then is not whether landscapes are capable of neutralizing acid rain, clearly they can do so, but rather how fast can they do it, and in a related context how long can they sustain it?

Initially, qualitative answers to these questions have been offered by geographically outlining "sensitive areas" regarding acid rain [2]. To elaborate on this further, however, a large body of chemical theory and geologic background can be called on to explain just why some areas are more sensitive to acid rain than others. This discussion will use the first principles of solution chemistry and crystal chemistry to describe and explain just how acid rain is neutralized by the geologic materials that make up the earth's surface.

BEDROCK AS A CHEMICAL AGENT

Throughout geologic history one of the principal functions of the geologic materials at the earth's surface has been to consume atmospheric acid, which normally has been carbonic acid [3]. The earth's surface in fact has acted as such an acid sink for billions of years. So, the advent of contemporary acid rain at the earth's surface is not an entirely new or unfamiliar phenomenon, but rather represents a change in an old, normal process. The one acid sink common to all landscapes is its bedrock and over the long term, the disposition of acid rain must eventually revert to this acid sink. The process by which this ultimate acid sink works is through the process called hydrolysis weathering [4].

Although the earth's crust contains every natural element in the periodic table, only eight elements (O, Si, Al, Fe, Ca, Na, K and Mg) are quantitatively important (Table I). This greatly simplifies understanding the geochemical processes operating on a worldwide basis. For example, we can characterize the bulk chemical behavior of crustal rocks (Table I) as that of a complex basic salt; that is, a compound of a weak acid (silicic acid) with a combination of strong bases. Symbolically, this characteristic can be represented as a chemical compound whose general formula is $MAlSiO_n$ where M is Ca, Na, K, Mg or Fe or some combination thereof [4].

From this generalized chemical perspective, then, the neutralization of acid rain can be seen as a reaction between the solid phases of the earth's crust and the acid rain itself

$$MAlSiO_n + H^+ = M^+ + Al^{3+} + H_4SiO_4^0$$

where H^+ represents SO_x and/or NO_x acidity. Considering the great mass of the earth's crust, this reaction implies that the potential acid-neutralizing capacity of the average land surface is enormous, involving billions of base equivalents per square meter of land surface.

The word "potential" is used advisedly because the generalized reaction described above is commonly a very slow one, even given a geologic time frame. So, although the acid-neutralizing capacity of crustal rocks may be great as well as thermodynamically possible, it may not happen very rapidly because of kinetic factors [5].

Table I. Elemental Composition of Average Crustal Rocks

Element	*Weight Percent*
O	46.60
Si	27.72
Al	8.13
Fe	5.00
Ca	3.63
Na	2.83
K	2.59
Mg	2.09
Σ	98.59

As suggested by Table I, the most likely minerals against which acid rain should impinge on the average are silicate minerals. Most silicate minerals in the earth's crust were formed under high temperature conditions, which do not generally pertain at the earth's surface. In other words, most silicate minerals at the earth's surface were not formed there, and are not thermodynamically stable there [5]. Metastability is then the rule, not the exception for silicate minerals at the earth's surface. This metastability is a powerful testament to the sluggish nature of silicate reactions at low temperature. Some silicate minerals have persisted in an unstable configuration for billions of years. We may draw another conclusion from this prevalent metastability: chemical *kinetics* are the relevant issue in low-temperature silicate reactions, and not necessarily chemical *thermodynamics* alone. This includes the hydrolysis weathering reaction through which silicate minerals may neutralize acid rain. Some silicate minerals react with acids more rapidly than others; some silicate minerals do not react with acids at all. These wide variations in acid reactivity are explained by the intrinsic crystal chemical properties of each mineral.

The following section will review the basic principles that govern the character and speed of silicate reactions with water and acids. It should be reiterated here that silicate minerals and the rocks formed from them are not renowned for their ability to dissolve in water or acids. On the contrary, to the layman and geologist alike, items made from silicate rocks such as "slate roofs," "granite monuments" and "soapstone chemical benches" connote the very opposite impression, i.e., one of permanence and indestructibility. Nevertheless, these materials do dissolve in water and acid at a rate that, although imperceptible by human standards, is important to the overall acid rain neutralization problem.

Silicate Hydrolysis

The reactivity of a silicate mineral to water or acid is largely a function of the silicate polymerization inherent to the mineral structure. The primary structural unit in silicate minerals is the SiO_4 tetrahedron, which may polymerize in various ways through a Si–O–Si covalent bond. Silicate minerals are conveniently classified on the basis of their silicate polymer configuration (Table II).

Generally, the more Si–O–Si linkages present in a mineral, the more difficult it is to dissolve the mineral in water or acid. We can see from Table II that the most water-soluble minerals are those at the top of the list, and those most insoluble are at the bottom.

The dissolution of a silicate mineral with water is conventionally called "hydrolysis weathering" or simply "weathering." In strict chemical terms hydrolysis is the dissolution of a salt in water, which produces an excess of H^+ or OH^-. Because most minerals that make up the earth's crust are basic salts, their hydrolysis leads generally to an excess of OH^-. However, there is one notable exception to this rule. The hydrolysis of pure quartz yields a weak acid, silicic acid, as an end product, which imparts a slight acid character to the reaction

Table II. Classification of Silicate Minerals

SiO_2 Polymer Configuration	*Percent Si–O–Si Linkages*	*Anion Formula*	*Si/O*	*Example*
Isolated Tetrahedra	0	SiO_4	1/4	Mg_2SiO_4 (forsterite)
Tetrahedra Doublet	25	Si_2O_7	1/3.5	$Ca_2Al_3O(SiO_4)(Si_2O_7)OH$ (epidote)
Single Chain	50	SiO_3	1.3	$MgSiO_3$ (enstatite)
Closed Chain	50	$(SiO_3)_n$	1.3	$Be_3Al_2(SiO_3)_3$ (beryl)
Double Chain	69	Si_4O_{11}	1/2.75	$Ca_2Mg_5(Si_4O_{11})_2(OH)_2$ (tremolite)
Sheet	75	Si_4O_{10}	1/2.5	$Mg_3(Si_4O_{10})(OH)_2$ (talc)
Framework	100	SiO_2	1/2	SiO_2 (quartz)

$$\underset{\text{(quartz)}}{SiO_2} + 2HOH = \underset{\text{(silicic acid)}}{H_4SiO_4^0} \qquad (1)$$

$$\log K^\circ = -4.00$$

In this and subsequent reactions, the equilibrium constants are taken or derived from the data of Lindsay [6].

It is significant that the quartz structure represents the highest possible level of silica polymerization, i.e., a Si:O ratio of 1:2 (Table II). The hydrolysis of quartz requires that a three-dimensional Si–O–Si network be broken down into a set of discrete SiO_4 tetrahedra. This depolymerization reaction is bimolecular, involving the breaking of a strong Si–O–Si bond and the simultaneous presence of an anion, in this case OH^-, to complete the new bonding arrangement. Note that this depolymerization process uses atomic oxygen, or its proxy OH^-, to convert the SiO_2 formula unit of quartz into the SiO_4 formula unit of silicic acid. In other words the Si:O ratio of the reactant is 1:2, while the Si:O ratio in the product is 1:4. OH^- ion is the vehicle through which this oxygen transfer is accomplished. It follows then that the amount of quartz dissolution increases directly with OH^- concentration

$$\underset{\text{(quartz)}}{SiO_2} + OH^- + HOH = SiO(OH)_3^- \qquad (2)$$

$$\log K^\circ = 0.29$$

where $SiO(OH)_3^-$ is the first ionization state of silicic acid. Comparing these reactions (Equations 1 and 2) shows that at equilibrium substantially more SiO_2 is dissolved in a basic regime than is dissolved in pure water.

Not only is the amount of SiO_2 dissolution increased by OH^-, but also the rate of its dissolution increases as well. This is shown by analogy with the depolymerization of silicic acid by bases

$$\underset{\text{(dimeric silicic acid)}}{(HO)_3Si\text{–}O\text{–}Si(OH)_3} + OH^- = \underset{\text{(monomeric silicic acid)}}{Si(OH)_4^0 + SiO(OH)_3^-} \tag{3}$$

In this reaction the fissioning rate, dP/dt, for Si–O–Si bonds has been shown to be proportional to OH^- concentration in a bimolecular reaction [7]

$$\frac{dP}{dt} = [P][OH^-]K_o \cdot \exp(-E/RT) \tag{4}$$

where P is the concentration of dimeric silicic acid

It is clear from the above that one of the most prevalent minerals found at the earth's surface, quartz, is not vulnerable to acids or acid rain. In fact, Equation 4 shows that weathering activity over a pure quartz terrane should be retarded by the acid rain phenomenon. From the viewpoint of acid rain neutralization, this has another important consequence. By virtue of its chemical properties, a pure quartz landscape, say one composed of quartz sand, has little acid-neutralizing capacity. Acid rain incident on such a landscape will end up as acid stream water, lakewater and groundwater, unless other independent neutralizing processes intervene. This indeed has been the case found in the quartz-sand area of New Jersey, an area affected by intense acid rain [8].

At the other extreme from quartz there are silicate minerals that are relatively reactive to acids and acid rain. A good example would be the mineral forsterite, which is abundant and widespread in certain volcanic terranes. The hydrolysis of forsterite is given by

$$\underset{\text{(forsterite)}}{Mg_2SiO_4} + 4HOH = 2Mg^{2+} + 4OH^- + H_4SiO_4^0 \tag{5}$$

$$\log K^\circ = -27.13$$

The magnitude of this reaction is obviously acid-dependent, which is shown by the reaction

$$\underset{\text{(forsterite)}}{Mg_2SiO_4} + 4H^+ = 2Mg^{2+} + H_4SiO_4^0 \tag{6}$$

$$\log K^\circ = 28.87$$

Note that for this reaction to take place no Si–O–Si bonds need to be broken as the silica configuration in forsterite, a SiO_4 anion, is already that of the silicic acid end product. In the hydrolysis of forsterite, then, only the relatively weak ionic bonds between Mg–O need to be broken. Although activation parameters for this reaction have not been worked out yet, it has been long recognized that forsterite is vulnerable to hydrolysis weathering [4]. That is, forsterite and minerals of like structure are the first to decompose in the weathering regime at the earth's surface. A landscape that contains abundant forsterite would thus be sensitive to acid rain in the sense that the bedrock would tend to weather more rapidly than normal. The other side of this coin, of course, is that such a landscape would also be able to neutralize the incident acid rain quickly.

The majority of silicate minerals, however, are somewhere between quartz and forsterite with respect to both their structural complexity and their hydrolysis properties. This is consistent with the fact that quartz and forsterite are at opposite ends of the silica polymerization spectrum (Table II) and most minerals are intermediate in their structural complexity. A good example is the mineral clinoenstatite, whose hydrolysis is described by

$$\underset{\text{(clinoenstatite)}}{MgSiO_3} + 3HOH = Mg^{2+} + 2OH^- + H_4SiO_4^0 \qquad (7)$$

$$\log K^\circ = -16.58$$

The structure of clinoenstatite is based on a chain of polymerized silica tetrahedra whose Si:O ratio is 1:3. During hydrolysis atomic oxygen must be provided to convert the SiO_3 unit of clinoenstatite into the SiO_4 unit of silicic acid. This oxygen is provided by the water molecule (Equation 7), or more specifically by the OH^- of ionized water. As shown previously OH^- is an aggressive reagent (Equation 2) and catalyst (Equation 3) for the fissioning of Si–O–Si bonds. On the other hand, in the hydrolysis of clinoenstatite there is also a clear need for H^+ to exchange for Mg^{2+} (Equation 7). It is important to note that in this reaction (Equation 7) there is a simultaneous demand for both an acid, H^+, and a base, OH^-. This demand is met, stoichiometrically at least, by the water molecule.

The hydrolysis of clinoenstatite produces an overall basic reaction (Equation 7), so the magnitude of the reaction is favored by the presence of H^+

$$\underset{\text{(clinoenstatite)}}{MgSiO_3} + 2H^+ + HOH = Mg^{2+} + H_4SiO_4^0 \qquad (8)$$

$$\log K^\circ = 11.42$$

Note, however, that H^+ cannot entirely replace water as a reagent as in the case of forsterite hydrolysis (Equations 5 and 6). There is no doubt that H^+ controls the magnitude of the reaction (Equations 7 and 8), but it is not clear in detail just how H^+ may affect the kinetics of the reaction. The activation parameters for this reaction are not known, but it seems most likely that the Si–O–Si depolymerization step is the rate-determining one, because of the strong covalent Si–O–Si bond that must be broken. If this is so, then the rate of clinoenstatite hydrolysis would be increased by a high OH^- concentration. We have a situation where on one hand the magnitude of the reaction is favored by high H^+ concentration, but on the other hand the speed of the reaction is favored by high OH^- concentration. So, there is a fundamental contradiction here between the demand for OH^- and H^- in the reaction and the coexistence of OH^- and H^+ in solution. Theoretically, a high concentration of OH^- combined with a high concentration of H^+ would optimize the amount of the hydrolysis taking place and presumably its kinetics, too. This can be shown by the theoretical reaction

$$\underset{\text{(clinoenstatite)}}{MgSiO_3} + 3H^+ + OH^- = Mg^{2+} + H_4SiO_4^0 \qquad (9)$$

$$\log K° = 25.42$$

This reaction can be considered an ephemeral, activated state and not really a mass action. It begs the question though: how does contemporary acid rain affect the weathering rate of minerals such as clinoenstatite? The answer is not known at the present time.

The discussion so far has been restricted to the reaction of single, pure mineral phases with water and acids. In nature, however, minerals rarely occur alone or in a pure state. Similarly, natural water is not a pure, inorganic solution. This latter factor has an important bearing on the acid rain neutralization problem. It has been pointed out that the presence of an anion was necessary to complete the fissioning of a Si–O–Si bond. OH^- is not the only, nor necessarily the best, anion to perform this function. F^- has the right size and charge to proxy for OH^-. F^- can coexist in solution with H^+, so that the optimum conditions specified in Equation 9 can be realized

$$\underset{\text{(clinoenstatite)}}{MgSiO_3} + 3H^+ + F^- = Mg^{2+} + SiF(OH)_3^0 \qquad (10)$$

Carried out to its logical end, a substantially different reaction takes place

$$\underset{\text{(clinoenstatite)}}{MgSiO_3} + 6H^+ + 4F^- = Mg^{2+} + SiF_4(g) + 3H_2O \qquad (11)$$

Standard practice in geochemistry takes advantage of Equation 11 to speedily dissolve silicate rocks and minerals in the laboratory, using HF spiked with H_2SO_4. Note, however, that the principle involved in Equation 11 is the same as in pure hydrolysis, i.e., an anion is needed for the silica depolymerization process, and H^+ is needed to exchange for the cation present.

The possible role of F^- in natural chemical weathering processes is thus intriguing, if unknown. The presence of small amounts of F^- in the weathering regime could possibly exert a powerful catalyzing effect on silicate hydrolysis. In this regard, if acid rain were composed of HF, instead of H_2SO_4 and HNO_3, its environmental impact would be profoundly different. With HF rain a silicate terrane would be just as susceptible to acid attack as, say, a limestone terrane. This illustrates the point that in the overall acid rain problem, it is not only the H^+ that matters, but also the composition of the acid anion.

Hydrolysis of Aluminosilicates

In Table II the various silicate anions in an unsubstituted form are described and the preceding section has described the hydrolysis properties of these pure silicates. However, in the double-chain, sheet and framework structures it is common for aluminum to systematically substitute for silicon. This substitution effects strong changes in the physical and chemical properties of the resulting structure and in some cases produces new minerals. A good example is the feldspar structure, which is a derivation of a SiO_2 polymorph structure. When one-quarter of the Si^{4+} in the SiO_2 structure, coesite, is replaced with Al^{3+}, a feldspar anion complex is formed

$$\underset{\text{(coesite)}}{4SiO_2^0} = \underset{\text{(coesite)}}{Si_4O_8^0} \longrightarrow \underset{\text{(feldspar anion)}}{AlSi_3O_8^-} \qquad (12)$$

When the aluminum substitution is coupled with the ordered addition of K, Na or Ca, a new mineral the feldspar structure results, e.g. $Na^+(AlSi_3O_8)^-$. Structurally then, the feldspars are framework structures (Table II) closely related to SiO_2. Chemically, however, the Al–Si substitution makes the feldspars radically different from the SiO_2 minerals to which they are related.

As a general rule, aluminum substitution tends to make a silicate structure more susceptible to hydrolysis and acid attack because each AlO_4 tetrahedron represents a weak link in the silica polymerization network. We may show this symbolically as the acid dissolution (not depolymerization) of an idealized SiO_4–AlO_4 anion complex

$$O_3Si\text{–}O\text{–}AlO_3^{7-} + 3H^+ = SiO_4^{4-} + Al^{3+} + 3OH^{3-} \qquad (13)$$

This reaction highlights the fact that H^+ is the active agent in the decomposition of a Si–O–Al bond. This stands in marked contrast to the Si–O–Si bond, where OH^- is the active agent during decomposition. Aluminosilicates are thus better agents of acid rain neutralization than pure, unsubstituted silicates. Generally, the more aluminum present the greater the acid solubility.

This relationship may be illustrated by comparing the acid solubility of a series of SiO_2 derivative structures. Note how acid dissolution potential varies with Al–Si substitution and/or Si:O ratio

$$\underset{\text{(coesite)}}{Si_4O_8} + 4HOH = 4H_4SiO_4^0 \qquad (14)$$

$$\log K^\circ = -12.20$$

$$\underset{\text{(albite)}}{Na(AlSi_3O_8)} + 4H^+ + 4HOH = Na^+ + Al^{3+} + 3H_4SiO_4^0 \qquad (15)$$

$$\log K^\circ = 3.67$$

$$\underset{\text{(anorthite)}}{Ca(Al_2Si_2O_8)} + 8H^+ = Ca^{2+} + 2Al^{3+} + 2H_4SiO_4^0 \quad (16)$$

$$\log K° = 26.10$$

Although it is definitely a framework silicate in structure (Table II) anorthite has the solution properties much like forsterite at the other end of the silica polymerization spectrum. That is, both are completely acid soluble, requiring no depolymerization step. Because of its aluminum substitution, anorthite has a Si:O ratio of 1:4, just like forsterite. From geologic evidence anorthite is known to be the most sensitive of the feldspars to hydrolysis weathering, albite is less so, and the silica polymorphs the least [4].

It follows that a landscape composed primarily of highly aluminous silicate rocks will be relatively sensitive to dissolution by acid rain and, conversely, such a terrane would be able to neutralize acid rain relatively quickly.

Stable Minerals at the Earth's Surface

So far this discussion has covered the chemical properties of the high-temperature, anhydrous, silicate minerals, which make up >90% of the earth's crust. As noted, these minerals are not in equilibrium with the environment of the earth's surface. When these minerals do hydrolyze at the earth's surface, they tend to produce a basic reaction, which can neutralize acid rain. Given the vast amounts of these minerals in the earth's crust, they represent an almost unlimited sink for disposing of acid rain over the long term. The difficulty here is that these minerals are not especially reactive at low temperature, so that their acid-neutralizing capacity is not readily accessible in the short term.

However, other common minerals in the earth's crust are water- and acid-reactive at low temperature in the short term. These minerals generally have been formed at the surface and are stable there. They are characteristic of the soil zone and certain sedimentary rocks and are conventionally called "secondary" or "sedimentary" minerals. Although these sedimentary minerals comprise only 8% of the earth's crust, they tend to be concentrated at or near the earth's surface. For example, about half of the surface of the North American continent is veneered with sedimentary rocks. So, even though they are much less abundant than the high-temperature silicate minerals overall, sedimentary minerals are a conspicuous presence on the earth's surface. Because they are positioned to contact acid rain and they tend to be acid-reactive, sedimentary minerals are of paramount importance in the short- and longer-term acid rain neutralization problem. In fact, whenever and wherever they are present, the sedimentary minerals are the preeminent solid agent of acid rain neutralization [2,9]. The origin of these sedimentary minerals, exemplified by the lime carbonates and various aluminum hydroxides, is closely tied to the hydrolysis of the silicates.

All the weathering reactions cited so far in this discussion have viewed weathering as a congruent solution process, i.e., all of the solid reactants are transformed into soluble components. There is strong empirical evidence showing that silicate hydrolysis does proceed as a congruent solution at least during its first step [5]. Under natural conditions, however, chemical weathering often leads to an incongruent decomposition process, depending on the supply of water, air and acid in the reaction. We have already shown how the feldspar anorthite dissolves congruently under acid conditions (Equation 16). However, if conditions are not too acid, anorthite hydrolysis may leave aluminum behind as a residual mineral in the soil zone

$$\underset{\text{(anorthite)}}{Ca(Al_2Si_2O_8)} + 2H^+ + 6HOH = Ca^{2+} + \underset{\text{(gibbsite)}}{2Al(OH)_3} + 2H_4SiO_4^0 \qquad (17)$$

$$\log K^\circ = 7.25$$

The mineral gibbsite and/or related aluminous minerals are characteristic of the so-called pedalfer soils, which typify much of the eastern half of the United States. If an old-age pedalfer soil, i.e., one with abundant gibbsite-like minerals present, should become subject to acid rain, the excess acidity would be absorbed quickly by the dissolution of gibbsite

$$\underset{\text{(gibbsite)}}{Al(OH)_3} + 3H^+ = Al^{3+} + 3HOH \qquad (18)$$

$$\log K^\circ = 8.04$$

This is a fast and reversible reaction. It is not, unfortunately, a permanent acid sink, because Al^{3+} itself behaves as a Bronsted acid. That is, it becomes a proton donor if $Al(OH)_3$ should be precipitated (see Equation 18). Although it may not be a permanent remedy for the neutralization of acid rain, soil $Al(OH)_3$ functions effectively as a short-term pH buffer over large regions that are affected by acid rain but are covered by pedalfer soils [10,11].

In climates that are arid and have alkaline conditions in the weathering zone, the hydrolysis of anorthite may take still another course

$$\underset{\text{(anorthite)}}{Ca(Al_2Si_2O_8)} + \underset{\text{(air)}}{CO_2} + 2HOH = \underset{\text{(calcite)}}{CaCO_3} + \underset{\text{(kaolinite)}}{Al_2Si_2O_5(OH)_4} \qquad (19)$$

$$\log K^\circ = 7.91$$

The weathering reaction in this case yields all insoluble products. Soils impregnated with calcite by this process are called pedocals. Pedocal soils cover much of the western half of the United States. An old-age pedocal soil represents a naturally "limed" soil, one capable of neutralizing acid rain promptly and totally.

$$\underset{\text{(calcite)}}{CaCO_3} + 2H^+ = Ca^{2+} + CO_2 + HOH \qquad (20)$$

$$\log K° = 9.74$$

For as long as its calcite reservoir holds out, a pedocal soil is then impervious to acidification through acid rain.

The soil clays, of which kaolinite is a common example (Equation 19), are also potential sinks for acid rain. Clays are acid-soluble minerals, but generally at acid levels substantially higher than that for calcite or gibbsite. For example the dissolution of kaolinite (Equation 17) is given by

$$\underset{\text{(kaolinite)}}{Al_2Si_2O_5(OH)_4} + 6H^+ = 2Al^{3+} + 2H_4SiO_4^0 + H_2O \qquad (21)$$

$$\log K° = 9.74$$

If the calcite and/or gibbsite reserves in a soil should become exhausted because of acid rain, then the clay minerals would be next in line to take over as an acid sink. Should the clay minerals become exhausted, the final recourse would be to the minerals of the bedrock basement.

The bedrock basement as we have noted is not necessarily composed of just silicate rocks and minerals. Sedimentary rocks, which are aggregates of clay minerals, carbonate minerals and various rock fragments, commonly occur at the earth's surface and as such represent an acid sink distinct from the soil zone. That is, clays and carbonates may be present in either or both the soil zone and its underlying sedimentary bedrock.

Whenever calcite or other carbonates coexist with silicate minerals, the calcite will preempt the role of acid rain neutralization. Even though the silicate mineral present could potentially act as an acid rain neutralizer, it will not be able to compete with the faster carbonate–acid reaction. As long as calcite is present at all in the soil zone or in bedrock, even in trace amounts, it will control the acid chemistry of the system. For example, a common sedimentary rock type is a quartz sandstone with a calcite cement. Though the quartz grains may be the dominant mineral present quantitatively, the calcite will be the dominant chemical control over any water in contact with it. This is a manifestation of the kinetic factor in the weathering process.

Perhaps because it is so reactive, many textbook discussions of chemical weathering use the dissolution of calcite (Equation 20) to exemplify the general weathering reaction. As we have seen, however, this is really a special case of weathering, an important one, but not typical overall. The popular perception that all chemical weathering is strictly an acid dissolution process may be attributed in some measure to this calcite paradigm. As we have seen, it is a useful simplification; however, if carried too far, it can be misleading about the real characterization of weathering. We need only point out that quartz monuments and buildings are immune to and indeed protected by the chemistry of our present acid rain.

Short-Term Acid Sinks

A landscape and its associated ecosystem have other means to neutralize acid rain besides the weathering of rocks and minerals. There are an assortment of ready acid sinks that interface between acid rain and the bedrock of the system. Chief among these are (1) the carbonate alkalinity dissolved in the ambient soil water, groundwater and stream water, (2) the base-exchange capacity of the local soils, and (3) the free basic cations contained in the system's biomass. These acid sinks are dynamically maintained within the system by expenditures of solar energy, generally through biologic and hydrologic activity. They tend to be open, steady-state systems, so that their reservoirs of free basic cations have to be sustained over the long term by a flux of new bases. Significantly, this supply of base comes ultimately from the underlying bedrock of the system by means of chemical weathering processes. The residence times for these base reservoirs varies widely from ecosystem to ecosystem, but usually ranges from a few months to a few decades. Because they are relatively fragile systems, delicately balanced between rate of base supply and loss, the intrusion of excess acids into the system may be reflected as a depletion or exhaustion of the normal pool of free bases. It has been recognized that "acidified waters," "depleted soils" and "degraded biomass" are the first signs of acid stress on an ecosystem [2].

Because of their easy access and quick response to acid rain, the waters, soils and vegetation of an ecosystem are the first acid sinks to be affected. They also tend to be the smallest, most sensitive and probably the most important of the various acid sinks; impinging as they do on the immediate life-support system. It must be borne in mind, however, there is a chain or progression of acid-base reactions that leads inexorably back to the ultimate, long-term acid sink of the landscape, the bedrock basement. The length, size and connections in this chemical chain determine exactly how a given landscape and ecosystem are going to accommodate the intrusion of any acid rain. The important point here is that over the long term, acid rain will inevitably affect and in turn will be affected by the geologic materials making up the landscape surface.

CASE STUDIES OF CHEMICAL WEATHERING RATE

To put the chemical effects of acid rain into perspective, a few case studies will be reviewed where the kind and amount of chemical weathering activity has been determined quantitatively. Cases where weathering proceeds under acid rain conditions and under normal rain conditions will be compared. To minimize the number of variables, only areas composed of granitic bedrock that have been glaciated will be considered. The case studies selected are:

- the South Cascade Glacier watershed of western Washington state [12];
- the Jamieson Creek watershed of southwestern British Columbia [13]; and
- the Hubbard Brook watershed of central New Hampshire [14].

The above areas also have in common a wet, cool and temperate climate.

South Cascade Glacier

The South Cascade watershed presently contains an active, temperate glacier [12]. The watershed supports no significant vegetation, except for lichens covering rocks and occasional sedges. The precipitation into the area is unpolluted and its acidity is controlled by atmospheric CO_2. The silicate rocks of this watershed are hydrolyzing at a rate of 0.9 eq-m^{-2}-y^{-1}. Carbonic acid is consumed in this process but is promptly replenished by the absorption of CO_2 from the atmosphere [12]. In effect H^+ is supplied to the hydrologic system on demand, in amounts that are essentially unlimited. Organic acids or biologic interactions are not factors in this system. One of the main controls over the kinetics of the weathering reaction here is the large area of rock surface that is produced by the grinding effect of the glacier. Fresh, chemically active crystal surfaces are continuously created by this glacial action [12].

Jamieson Creek

The Jamieson Creek watershed lies just northwest of the South Cascade Glacier in British Columbia [13]. The glaciers left this area some 10,000 years ago and the watershed is now covered by heavy, virgin coniferous forest. An organic rich pedalfer (podzol) soil also mantles the area. The Jamieson Creek area is presumably what the South Cascade area will look like in 10,000 years or so. The precipitation in the area has an average pH of 5.3 and is CO_2 controlled, analogous to the South Cascade situation. Presently, the hydrolysis weathering of the silicate rocks in the Jamieson Creek area is proceeding at a rate of 0.3 eq-m^{-2}-y^{-1}, or substantially less than at the nearby South Cascade Glacier area. The reason for this lesser weathering activity is probably due to the lack of fresh mineral surfaces in the local bedrock. Unlike the South Cascade situation, the rocks at Jamieson Creek are "aged" by 10,000 years and each rock surface is covered with a rime of weathered rock products. The rate of hydrolysis is thus diffusion controlled, i.e., water must diffuse in and out of the weathering rime before fresh silicate material can be contacted and hydrolyzed.

Hubbard Brook

The Hubbard Brook area shares a common history with Jamieson Creek in that glaciation left both areas some 10,000 years ago. In almost every other respect, the two areas are also similar: heavy forest, organic-rich soils (podzols), wet climate, and granitic bedrock [13,14]. The outstanding difference between the two areas is the presence of acid rain at Hubbard Brook (pH $\simeq$4.1) and its absence at Jamieson Creek (pH $\simeq$5.3).

The silicate rocks in the Hubbard Brook area are hydrolyzing at a rate of 0.2 eq-m^{-2}-y^{-1}, somewhat less than the rate for Jamieson Creek (0.3 eq-m^{-2}-y^{-1}). At face value at least, the weathering rate at Hubbard does not appear to be unusual, despite the flux of acid rain. However, it is unknown whether the onset of acid rain in the Hubbard Brook area has increased or decreased the chemical weathering rate.

Quantitatively then, the annual consumption of acid by the landscape at Hubbard Brook is not exceptional. This may be attributed to the replacement of the normal acids at Hubbard Brook, presumably carbonic and organic acids, by the SO_x and NO_x of acid rain. In effect acid rain has not added to the normal acidity of the system, but instead has replaced it altogether.

In its qualitative character the stream water at Hubbard Brook does have unusual properties, it is notably rich in dissolved aluminum [10]. The dissolved aluminum comes from inorganic reactions and is especially conspicuous in low-order (short pathlength) stream water [10]. The source of the dissolved aluminum is a gibbsite-like mineral in the soil zone, which dissolves in response to the high acid concentrations associated with acid rain [10,11]. One of the more important consequences of contemporary acid rain on a landscape such as Hubbard Brook is this mobilization, translocation and redeposition of soil aluminum [15]. The soils over the Northeast are thus being subtly but definitely altered by acid rain from their original, natural state [15,16].

The neutralization of acid rain in the Hubbard Brook area also takes an unusual course. It may be viewed as a two-step process [10]. In the first step, H^+ acidity from the atmosphere is largely converted into Al^{3+} acidity by the hydrolysis of a gibbsite-like mineral in the soil zone. Stream water at this state is a mixture of H^+ acidity and Al^{3+} acidity. The next step in the acid neutralization process occurs more slowly as the water makes its way downstream, and its residence time in the system (and pathlength) increases. The basic cations released by silicate hydrolysis are added slowly as a function of water pathlength and/or residence time [10]. Along with the basic cations, silica is also systematically added to the stream water. As we have seen, both basic cations and $H_4SiO_4^0$ are by-products of silicate mineral hydrolysis. Their joint, synchronized appearance in Hubbard Brook testifies that silicate mineral hydrolysis is taking place in the system. This second step, silicate mineral hydrolysis, is the final one in the acid neutralization process, both H^+ and aluminum acidity in this system being permanently titrated away by the basic cations produced. The ultimate acid sink then in the Hubbard Brook system is its silicate bedrock.

If the comparisons between the South Cascade Glacier, Jamieson Creek and Hubbard Brook watersheds are at all valid, we can conclude that acid rain does not necessarily elicit a quantitative chemical effect so much as a qualitative one. Both quantitative and qualitative aspects can be explained by simple chemical principles. That is, H^+ is not the only nor even the most important rate-determining factor in hydrolysis weathering (see Equations 9 and 10). On the other hand, H^+ is indeed the crucial factor in the dissolution of gibbsite (see Equation 18). Perhaps the main point to be learned from all three case studies, however, is that silicate minerals, through their hydrolysis, can and do provide the long-term sink for acids fluxing into the system—both natural acids and that from acid rain.

SUMMARY

The hydrolysis of most geologic materials produces an excess of OH^- in solution, which can serve as a sink for acid rain. For the earth's crust as a whole the hydrolysis reaction can be put in a generalized form [4]

$$\underset{\text{(earth's crust)}}{MSiALO_n} + H_2O = M^+ + OH^- + \underset{\text{(silicic acid)}}{[Si(OH)_{0\text{-}4}]_n} + \underset{\text{(alumina)}}{[Al(OH)_6]_n} \quad (22)$$

One conspicuous exception to this reaction is the common mineral quartz, whose hydrolysis produces an acid reaction. Landscapes composed exclusively of quartz, although rare, are thus precluded from being an acid sink at all. A good example of this is shown by the sand barrens of New Jersey [6].

The key to understanding hydrolysis weathering is chemical kinetics. This is because most minerals are silicates, and silicate reactions are very sluggish at low temperatures. In contrast, sedimentary minerals (exclusive of quartz) tend to be more reactive at low temperatures. Thus, the carbonates, clays and gibbsite-related minerals are "fast" sinks for acid rain. Terranes with an abundance of these materials will not be vulnerable to acid rain, at least with regard to their soil chemistry.

In contrast, terranes composed of only silicate minerals do not have the same intrinsic ability to rapidly neutralize acid rain. Silicate hydrolysis is a slow, complex reaction, being bimolecular at the very least with a high activation energy associated with the breaking of Si–O–Si covalent bonds. The highly polymerized silicate structures are the most difficult to hydrolyze; the silicates with no SiO_4 polymerization are the least difficult. Aluminum substitution for silicon in tetrahedral positions renders a silicate structure more vulnerable to hydrolysis or acid attack, the AlO_4 tetrahedra being a weak link in the polymerization network. OH^- and F^- anions are crucial reagents in the depolymerization of Si–O–Si bonds.

Extrinsic factors also affect the rate at which a given rock will weather and/or neutralize acid rain. The rate of all liquid–solid reactions is enhanced by increasing the surface area of the solid exposed to the reacting solution. Finely ground or defective crystals are thus more reactive than large, perfect crystals. Massive bedrock of any kind is similarly less reactive than its pulverized equivalent, all other factors being equal. For incongruent hydrolysis reactions the nucleation and growth of new minerals present a kinetic obstacle to the overall reaction. All reactions, including hydrolysis, are strongly affected by temperature, so hot climates will be inherently more reactive in their weathering activity than cold climates.

When considered over the long term, much of the earth's surface is characterized by an integrated network of chemical reactions. At one end of the chemical network are the geologic materials underlying the landscape, at the other end is the atmosphere. The soil or weathering zone represents the interface between the two. The atmosphere provides mostly volatile constituents to the soil system while geologic materials provide most of the nonvolatile components. Climate, hydrology and biology interact to provide the energy to drive the system. After geologic spans of time, some chemicals, including basic cations, may accumulate in the soil zone in the form of chemically metastable reservoirs. Some of these reservoirs include the organic matter of the soil zone, the base-exchange capacity of the soil zone and the carbonate alkalinity of soil water. Overall, this chemical network of the soil zone and its underlying bedrock presents a hierarchy of reactions that provide acid sinks. In order of *decreasing* availability in a kinetic sense these are:

1. the free-base pools of the soil zone, such as soil-water alkalinity, base exchange capacity of the soil and organic matter;
2. the secondary minerals in the soil zone, such as calcite, $Al(OH)_3$ and clays;
3. sedimentary minerals in the bedrock, such as carbonate and clay minerals;
4. the silicate minerals of the bedrock basement.

The presence of any acid sink higher in the hierarchy will preempt the action of acid sinks lower than itself. With the exception of deep silicate bedrock, i.e., acid sink 4 above, one or more of the acid sinks may be missing from any given landscape. It is important to note that the "sensitive areas" with respect to acid rain are precisely those missing the top echelon of acid sinks in this hierarchy. That is, landscapes with thin or no soils, no sedimentary minerals present, and a basement of massive silicate rock are most vulnerable to acid rain impacts. This aptly describes the contemporary situation in the Precambrian shield areas of Canada and Scandinavia and their well known acid rain effects.

For purposes of predicting how a given landscape will respond to acid rain, a look at a geologic map and a soil map of the area concerned will provide some powerful clues. Areas with deep pedocal soils, finely pulverized bedrock or sedimentary bedrock will most likely accommodate acid rain rather well—at least over the short term. For a deeper understanding an appraisal of the exact mineral assemblages in the bedrock may be called for.

Throughout geologic history the earth's landscape has behaved as an acid sink for atmospheric acids, mainly carbonic acid. The acid–base reactions between rain and various geologic materials can be described and explained by simple chemical principles and equations. The strong acids of contemporary acid rain have the potential to either slow down or speed up the rate of local hydrolysis weathering, depending on the intrinsic chemical makeup of the affected terrane. The evidence so far, although anecdotal in scope, suggests that acid rain has not greatly changed the amount of hydrolysis weathering taking place overall. However, the same evidence shows that the kind of hydrolysis taking place has been altered substantially. If the acid concentration of acid rain should continue increasing, then we can predict from theory further qualitative changes in the chemical weathering of geologic materials.

ACKNOWLEDGMENTS

This is a contribution of the Hubbard Brook Ecosystem Study. Partial support was also provided by NSF grant EAR-8025719.

REFERENCES

1. Johnson, N.M., R.C. Reynolds and G.E. Likens. "Atmospheric Sulfur: Its Effect on the Chemical Weathering of New England," *Science* 177:514–516 (1972).
2. Likens, G.E., R.F. Wright, J.N. Galloway and T.J. Butler. "Acid Rain," *Sci. Am.* 241:43–50 (1979).
3. Garrels, R.M., and F.T. Mackenzie. *Evolution of Sedimentary Rocks* (New York: W.H. Norton, 1971), pp. 127–128.
4. Keller, W.D. *The Principles of Chemical Weathering* (Columbia, MO: Lucas Bros., 1957), p. 111.
5. Berner, R.A. "Kinetics of Weathering and Diagenesis," *Rev. Mineral.* 8:111–134 (1981).
6. Lindsay, W.L. *Chemical Equilibria in Soils* (New York: John Wiley & Sons, Inc., 1979), p. 449.
7. Nemecz, E. *Clay Minerals* (Budapest: Akademiai Kiado, 1981), pp. 379–394.
8. Johnson, A.H. "Acidification of Headwater Streams in the New Jersey Pine Barrens," *J. Environ. Qual.* 8:383–386 (1979).
9. Norton, S.A. "Geologic Factors Controlling the Sensitivity of Aquatic Ecosystems of Acidic Precipitation," in *Atmospheric Sulfur Deposition,* D.S. Shriner, et al., Eds. (Ann Arbor, MI: Ann Arbor Science Publishers, Inc., 1980).
10. Johnson, N.M., C.T. Driscoll, J.S. Eaton, G.E. Likens and W.H. McDowell. " 'Acid Rain', Dissolved Aluminum and Chemical Weathering at the Hubbard Brook Experimental Forest, New Hampshire," *Geochim. Cosmochim. Acta* 45(9):1421–1437 (1981).
11. Driscoll, C.T. "Chemical Characterization of Some Dilute Acidified Lakes and Streams in the Adirondack Region of New York State," PhD Thesis, Cornell University (1980).
12. Reynolds, R.C., and N.M. Johnson. "Chemical Weathering in the Temperate Glacial Environment of the Northern Cascade Mountains," *Geochim. Cosmochim. Acta* 36:537–554 (1972).
13. Zeman, J.L. "Chemistry of Tropospheric Fallout and Streamflow in a Small Mountainous Watershed Near Vancouver, British Columbia," PhD Thesis, University of British Columbia (1973).
14. Likens, G.E., F.H. Bormann, R.S. Pierce, J.S. Eaton and N.M. Johnson. *Biogeochemistry of a Forested Ecosystem* (New York: Springer-Verlag, 1977).
15. Parnell, R. "Chemical Weathering and Aluminum Cycling in Subalpine and Alpine Soils and Tills, Mt. Moosilauke, N.H.," PhD Thesis, Dartmouth College (1981).
16. Cronan, C.S., N.A. Reiners, R.C. Reynolds and G.E. Lang. "Forest Floor Leaching: Contributions from Mineral, Organic and Carbonic Acids in New Hampshire Subalpine Forests," *Science* 200:309–311 (1978).

CHAPTER 4

Aluminum Speciation and Equilibria in Dilute Acidic Surface Waters of the Adirondack Region of New York State

Charles T. Driscoll
Joan P. Baker
James J. Bisogni
Carl L. Schofield

Recent studies of dilute (low-ionic-strength) lakes, streams and groundwaters in the northeastern United States [1–3], southern Norway [4] and Sweden [5] have shown that acidic deposition readily mobilizes aluminum from soil to the aquatic environment. Although this fact is well documented, detailed investigations of aqueous aluminum chemistry in natural acidified waters are limited.

Dissolved monomeric aluminum occurs as aquoaluminum, as well as hydroxide, fluoride, sulfate and organic complexes [6]. Past investigations of aluminum in dilute natural waters have often ignored nonhydroxide complexes of aluminum [2,7,8]. Johnson et al. [3] investigated spatial variations in aluminum chemistry within a headwater stream in New Hampshire. While total monomeric aluminum and hydrogen ion levels decreased with decreasing elevation, aluminum speciation also shifted. Aquoaluminum and aluminum hydroxide complexes decreased substantially with decreases in elevation. Aluminofluoride forms remained constant throughout the experimental reach. Aluminoorganic forms increased in concentration with decreasing elevation. A comparison of aquoaluminum levels with theoretical values based on the solubility of various aluminum minerals suggested that aqueous aluminum levels were regulated by aluminum trihydroxide [$Al(OH)_3$].

Aqueous aluminum is significant because it is a weak acid/base system and therefore an important buffer in dilute acidic waters. The potential pH buffering of aluminum has been discussed by several investigators [9–11].

Elevated levels of aqueous aluminum appear to have deleterious effects on fish in dilute waters [2,12]. Fish fry are particularly sensitive to hydrolyzed aluminum [13]. Therefore, pH, natural organic and inorganic (F^-, SO_4^{2-}) ligands significantly influence aluminum toxicity [13,14]. Elevated levels of soluble ligands restrict aluminum hydrolysis and mitigate aluminum toxicity to fish.

The intent of this investigation was to describe alumino–ligand interactions in acidic aquatic ecosystems. To achieve this end, water quality data were collected from several acidic lakes and streams in the Adirondack Mountain region of New York state. These data together with chemical equilibrium calculations were used to assess the distribution of dissolved aluminum species in these waters.

EXPERIMENTAL METHODOLOGY

Three lakes located in the southwestern region of the Adirondack Park, North Lake, Big Moose Lake and Little Moose Lake, and several of their tributaries (a total of 24 sites) were sampled 15 times over the course of a year (August 1977 to August 1978) [Table I]. The areas surrounding all three lakes have a similar geologic composition [15] and are largely underlain by silicate bedrock. Water samples were collected approximately every three weeks, in the spring, summer and fall, and every four weeks in the winter. At certain times of the year (late winter and spring, when ice was forming and melting) some sites were not accessible and therefore were not sampled. Sediments were collected from a tributary to Big Moose Lake, West Pond Outlet, Big Moose Lake, North Lake and Little Moose Lake in June 1978.

Water samples were collected in 1-L polypropylene bottles and were centrifuged at 7000 g for 20 min to achieve separation of solid particles. pH was determined potentiometrically with a glass electrode. Acid-neutralizing capacity was determined by strong acid titration to an endpoint calculated by a Gran plot analysis [16]. Free fluoride was measured potentiometrically with a fluoride ion-selective electrode [17]. Sulfate was measured by the turbidimetric method [18].

Table I. Sampling Site Information

	Big Moose Lake	*North Lake*	*Little Moose Lake*
Location	74°50′ W, 43°30′ N	74°55′ W, 43°43′ N	74°55′ W, 43°42′ N
Surface Area (ha)	513	161	272
Elevation (m)	556	557	545
Maximum Depth (m)	22	16	37
Lake Sampling Sites	7	8	1
Tributary Sampling Sites	3	4	1

Total organic carbon (TOC) was determined by oxidation followed by infrared detection of CO_2. Basic cations (Ca^{2+}, Mg^{2+}, Na^+, K^+) were determined by atomic absorption spectrophotometry (AAS). In the determination of calcium and magnesium, lanthanum was added to samples to minimize potential interference by silica, aluminum, sulfate or orthophosphate [19]. Specific conductance was determined by using a conductivity bridge [18]. Dissolved silica was measured using the molybdosilicate method [18].

Three forms of aluminum were measured. Total aluminum was analyzed by a colorimetric ferron-orthophenanthroline method after samples were acid digested for 0.5 h [20]. The magnitude of total aluminum measurements was checked against graphite-furnace AAS, and close agreement with the colorimetric procedure was observed. Monomeric aluminum was measured using the same procedure without acid digestion, as suggested by Smith [21]. Organic monomeric aluminum was separated using an ion exchange chromatography technique and measured as monomeric aluminum [9]. Inorganic monomeric aluminum was calculated as the difference between monomeric aluminum and organic monomeric aluminum. Acid-soluble aluminum (e.g., particulate, colloidal Al) was also calculated as the difference between total aluminum and monomeric aluminum.

Water quality data were examined with respect to chemical equilibria. The thermodynamic relationships used in this study are summarized in Table II [22–30]. The equations for calculating various aluminum forms are summarized elsewhere [3,9]. Activity corrections were made utilizing the Debye–Huckel relationship [26].

X-ray diffractometer tracings were made on sediments to check for the presence of solubility-controlling minerals. These tracings were done with a copper target using a nickel filter.

RESULTS

General Observations

The composition of the dilute Adirondack waters sampled differed from that of typical freshwaters [31]. In most freshwaters, the total dissolved solids content is dominated by basic cations (Ca^{2+}, Mg^{2+}, Na^+, K^+) and bicarbonate. In Adirondack waters, calcium was generally the dominant cation; however, acidic cations (H^+, Al) comprised a high percentage of total solutions cations on an equivalence basis (Figure 1). Above pH 5.5, the cation chemistry was dominated by basic cations. Below pH 5.5, acidic cations comprised a significant amount of the solution cation chemistry. Bicarbonate levels were low in these waters because of the generally low pH and limited carbonic acid weathering. In Adirondack surface waters, a high portion of the mass of total dissolved solids was attributed to dissolved silica, sulfate and natural organic carbon.

Table II. Equilibrium Relationships Used in This Study

Equation	*Equilibrium Constant*	*Reference*
Hydroxide Ligands		
$Al^{3+} + H_2O = Al(OH)^{2+} + H^+$	1.03×10^{-5}	22
$Al^{3+} + 2H_2O = Al(OH)_2^+ + 2H^+$	7.36×10^{-11}	22
$Al^{3+} + 4H_2O = Al(OH)_4^- + 4H^+$	6.93×10^{-23}	22
Fluoride Ligands		
$Al^{3+} + F^- = AlF^{2+}$	1.05×10^{7}	23
$Al^{3+} + 2F^- = AlF_2^+$	5.77×10^{12}	23
$Al^{3+} + 3F^- = AlF_3$	1.07×10^{17}	23
$Al^{3+} + 4F^- = AlF_4^-$	5.37×10^{19}	23
$Al^{3+} + 5F^- = AlF_5^{2-}$	8.33×10^{20}	23
$Al^{3+} + 6F^- = AlF_6^{3-}$	7.49×10^{20}	23
Sulfate Ligands		
$Al^{3+} + SO_4^{2-} = AlSO_4^+$	1.63×10^{3}	24
$Al^{3+} + 2SO_4^{2-} = Al(SO_4)_2^-$	1.29×10^{5}	24
Aluminum Trihydroxide		
$Al(OH)_3 + 3H^+ = Al^{3+} + 3H_2O$		
Synthetic Gibbsite	1.29×10^{8}	22
Natural Gibbsite	5.89×10^{8}	22
Microcrystalline Gibbsite	2.24×10^{9}	25
Amorphous Aluminum Trihydroxide	6.31×10^{10}	26
Aluminum Silicates		
$\frac{1}{2} Al_2Si_2O_5(OH)_4 + 3H^+ = Al^{3+} + H_4SiO_4 + \frac{1}{2} H_2O$		
Kaolinite	2.00×10^{3}	26
Halloysite	4.37×10^{5}	27
Variable-Composition Aluminum Silicates		
$Al(OH)_{3(1-x)} SiO_{2x} + (3 - 3x)H^+ = (1 - x)Al^{3+} + xH_4SiO_4 + (3 - 3x)H_2O$[a]		
Amorphous Ideal Aluminosilicate	$10^{-5.7} + 1.68pH$	28
Reversible Nonideal Aluminosilicate	$10^{-5.89} + 1.59pH$	28
Alunite		
$KAl_3(SO_4)_2(OH)_6 = K^+ + 3Al^{3+} + 2SO_4^{2-} + 6OH^-$	3.98×10^{-86}	29
Jurbanite		
$Al(SO_4)(OH)\ 5H_2O = Al^{3+} + SO_4^{2-} + OH^- + 5H_2O$	1.58×10^{-18}	30

[a] Where $x = 1.24 - 0.135pH$.

Ionic Strength and Specific Conductance

Ionic strength (I) is an important aspect of aqueous chemistry that can alter ionic equilibria [26].

$$I = 1/2 \Sigma C_i Z_i^2$$

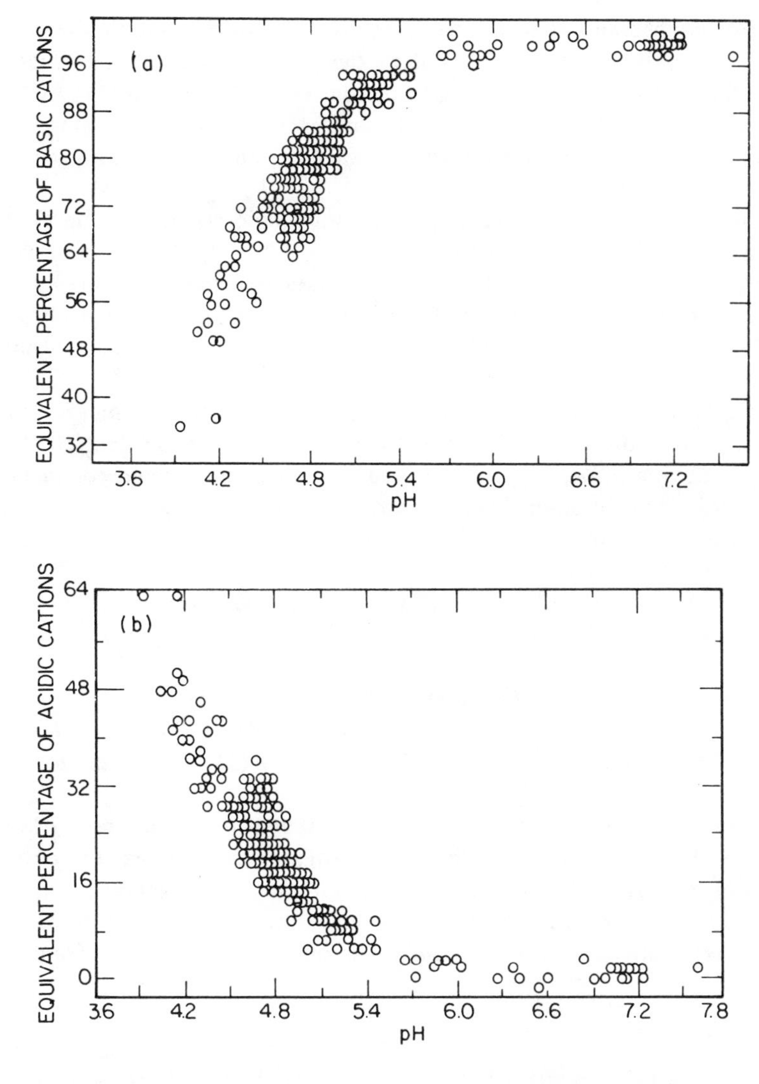

Figure 1. Percentage of solution (a) basic (Ca^{2+}, Mg^{2+}, Na^{+}, K^{+}) and (b) acidic (H^{+}, Al) cation composition on an equivalence basis as a function of solution pH. Acidic cations comprise a significant fraction of total solution cations below pH values of 5.5. Each data point represents an individual sample from this study.

where I = solution ionic strength
C_i = molal concentration of a particular constituent
Z_i = charge of a particular constituent

Chloride and nitrate, which are major constituents contributing to ionic strength, were not measured in this study. To estimate ionic strength, the chloride and nitrate contribution was calculated by the difference in solution cationic and anionic charge. It is noteworthy that in these calculations it was necessary to consider the contribution of individual aluminum species. For example, Al^{3+}, AlF^{2+} and $Al(OH)_2^+$ have a different charge and solution concentration; therefore, thermodynamic calculations must be made to estimate individual aluminum species contribution to ionic strength.

Adirondack waters are very dilute. Mean ionic strength value ($I \pm P = 0.05$) for total, lake and stream data sets were $3.2 \pm 1.1 \times 10^{-4}$, $3.1 \pm 0.9 \times 10^{-4}$ and $3.4 \pm 1.2 \times 10^{-4}$, respectively. It is apparent that ionic strength values were very low and there was little difference between the ionic strength of Adirondack lakes and streams. As a result, ion activity was very close to ion concentration in value.

Specific conductance values of Adirondack waters increased significantly with decreases in pH values. This is due to the relatively high equivalent conductance value of hydrogen ion (349.8 mho-cm²-eq⁻¹). When the specific conductance of hydrogen ion was subtracted from measured specific conductance values, a significant correlation was observed with ionic strength

$$I = 7.04 \times 10^{-6} \times \{K - ([H^+] \times \lambda_o^{H+} \times 1000^{-1})\} + 1.37 \times 10^{-4}$$
$$r^2 = 0.46;\ P < 0.0001$$

where K = measured specific conductance (μmho-cm^{-1})
$[H^+]$ = sample hydrogen ion concentration (*M*)
λ_o^{H+} = equivalent conductance of hydrogen ion (349.8 mho-cm^2-eq^{-1} at 25C)

This empirical relationship is somewhat different from that observed by Lind ($I = 1.6 \times 10^{-5}$ K) [32], although this investigator showed that the correlation varied considerably with composition of major ions. The relationship developed in our study is useful because it is relatively difficult to measure all of the parameters needed to compute ionic strength, while it is relatively easy to determine specific conductance.

Aluminum Chemistry

Aluminum levels in Adirondack surface waters were extremely variable. Examination of total aluminum levels as a function of pH provided little insight as to the response of this element to hydrogen ion inputs and acidification (Figure 2a). Data points in Figure 2a represent all samples collected during the study and involve a variety of sampling dates and sites. It is apparent that data were considerably scattered. The general trend, however, was an increase in total aluminum levels with decreases in pH below 5.0.

A better understanding of aluminum interactions in natural waters was obtained when the concentrations of the different aluminum fractions were examined relative to chemical changes in the aquatic environment (Figures 2b and 3). At

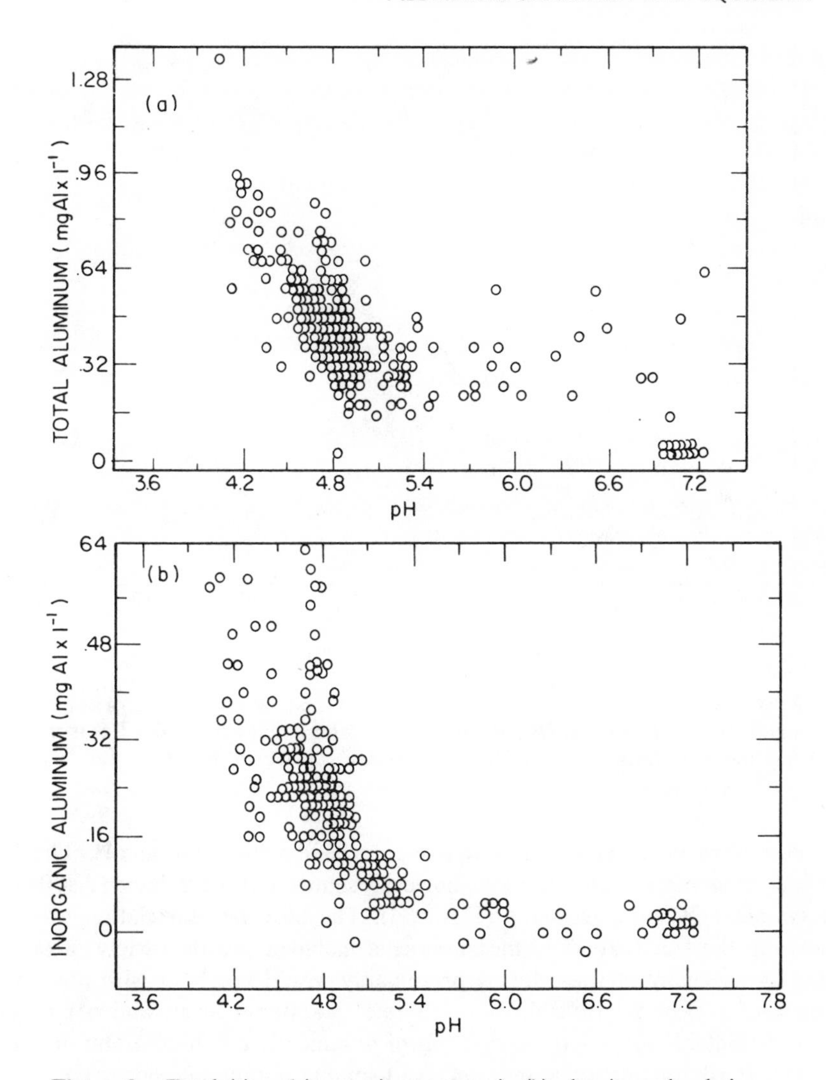

Figure 2. Total (a) and inorganic monomeric (b) aluminum levels in Adirondack surface waters as a function of solution pH. Total aluminum values were scattered, while monomeric aluminum values exhibited an exponential increase with decreasing solution pH. Each data point represents an individual sample from this study.

neutral pH values, inorganic monomeric aluminum levels were low. As the solution pH decreased, the inorganic monomeric aluminum concentration increased exponentially (Figure 2b). This trend is similar to what would be anticipated from an aluminum trihydroxide solubility model.

Concentrations of organic monomeric aluminum, on the other hand, appeared to be independent of pH. Neither organically complexed aluminum nor TOC mea-

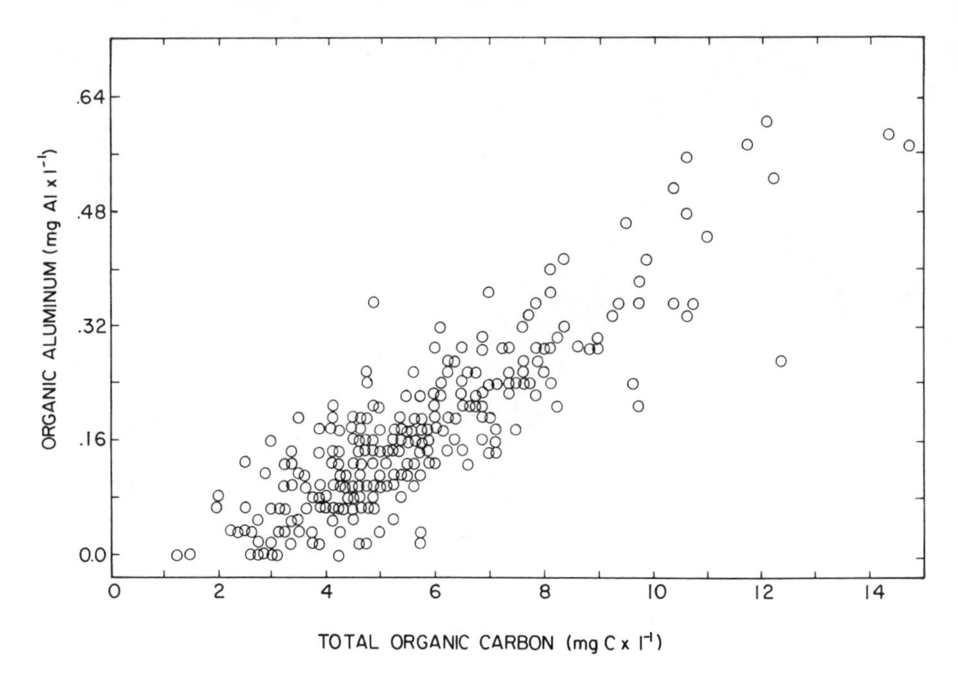

Figure 3. Organic monomeric aluminum levels in Adirondack surface waters as a function of solution TOC concentration. Organic monomeric aluminum exhibited a linear increase with increasing TOC (Al-Org = $-0.084 + 0.046$TOC, $r^2 = 0.76$, $P < 0.0001$).

surements were found to be significantly correlated with solution pH. The level of organic monomeric aluminum was, however, significantly correlated ($P < 0.0001$) with sample TOC measurements (Figure 3). The observed correlation was high considering the fact that TOC measurements included a wide variety of organic carbon forms; many of these forms presumably would not bind with aluminum. The lack of a clear relationship between total aluminum levels and pH may be largely attributable to variations in levels of organically complexed aluminum.

The third fraction of aluminum, acid-soluble aluminum, occurred only in low concentrations. At any given site, the concentration of acid-soluble aluminum was relatively constant over the course of the study. Levels of acid-soluble aluminum were found to not correlate with either pH or TOC measurements.

Temporal fluctuations in aluminum level and form in Adirondack waters also reflected temporal variations in pH and organic carbon levels (Figure 4). Levels of inorganic aluminum increased in streams during rainfall and snowmelt events, when waterflow was high and pH values were low. During low-flow, neutral-pH conditions in streams, inorganic monomeric aluminum levels were very low. During the summer months, when microbial decomposition activity was greatest and podzolization was intense, levels of TOC increased in stream systems. Likewise, levels of organically complexed aluminum increased.

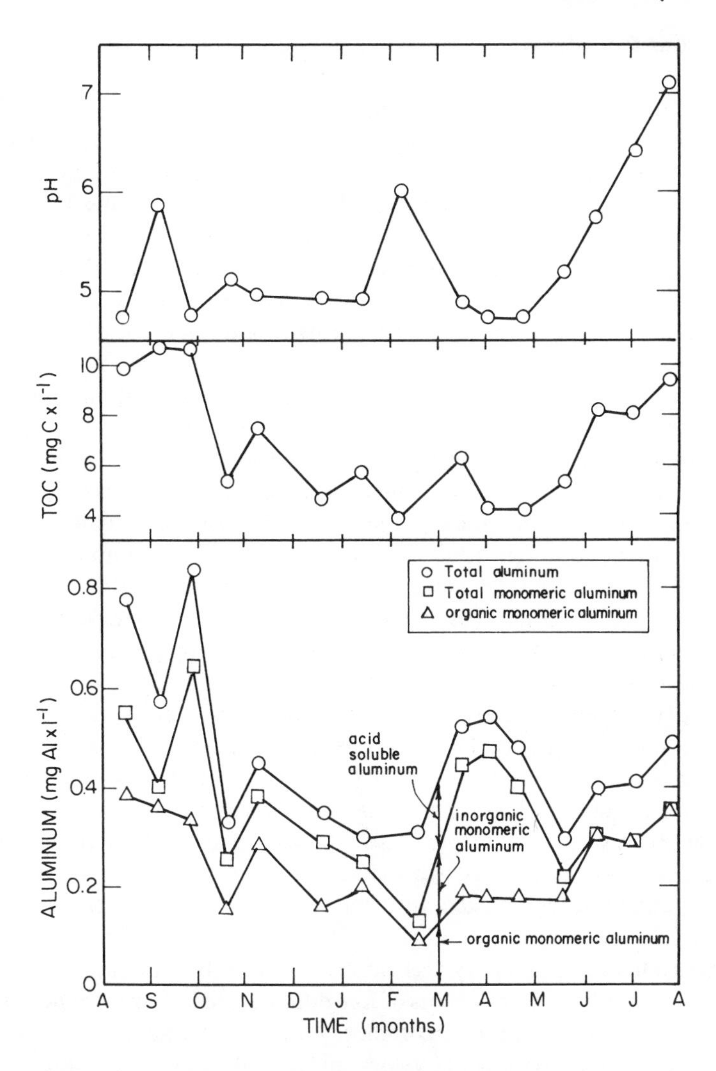

Figure 4. Temporal changes in pH, TOC and aluminum forms for Canachagala Creek from August 1977 to August 1978. Low pH and high levels of inorganic aluminum were observed after rainfall (August 15, September 27) and during snowmelt (March 15, April 2 and April 23). Low flow conditions produced elevated pH values and decreased inorganic aluminum values. TOC and organic aluminum were generally high in the autumn and summer and low in the winter and spring. ○ = total aluminum; □ = total monomeric aluminum; △ = organic monomeric aluminum.

Aqueous Coordination Chemistry of Aluminum

Soluble aluminum forms strong complexes with hydroxide, fluoride, sulfate and natural organic matter [6,33]. To compute the concentrations of various inorganic aluminum forms, inorganic monomeric aluminum, free fluoride, pH and sulfate measurements, and thermodynamic equilibrium constants (Table II) were used. All of these forms of dissolved aluminum were present in Adirondack surface waters to some degree. To illustrate the relative magnitude of aluminum forms, a tabulation is presented for all data, lake and stream (Table III). Monomeric organic complexes were the predominant forms of aluminum in our study accounting for, on the average, 37% of the total aluminum and 44% of the monomeric aluminum. Aluminum fluoride complexes were the second major form of aluminum and the predominant form of inorganic aluminum, accounting for 25% of the total aluminum and 29% of the monomeric aluminum. Aquoaluminum and soluble aluminum hydroxide complexes were generally less significant than aluminum fluoride complexes. Aluminum sulfate complexes were small in magnitude.

Both aluminum sulfate ($Al–SO_4$) (Figure 5a) and aluminum fluoride (Al–F) (Figure 5b) complexes were more significant at lower pH values and decreased in magnitude with increasing pH. However, there was a distinct difference in the functional dependence of the two aluminum forms with pH. The logarithm of the concentration of aluminum complexed with sulfate increased linearly with decreasing pH. Sulfate-complexed aluminum represented only a small percentage of the total solution sulfate. Therefore, as aquoaluminum levels increased with decreasing pH, so did sulfate-complexed aluminum. The magnitude of fluoride complexed aluminum reached a maximum value at about pH 5.5 and then leveled off at lower pH levels. In contrast to sulfate, at low pH values almost all of the aqueous fluoride was complexed with aluminum. Therefore at low pH values the quantity of aluminum complexed with fluoride was limited by the total fluoride concentration of the solution.

Aqueous fluoride chemistry was largely regulated by aluminum and hydrogen ion levels in Adirondack waters. Plots of free fluoride, aluminum-complexed fluoride and the sum of free and aluminum-complexed fluoride as a function of pH illustrate this concept (Figure 6). At low pH values ($pH < 5.5$), solution fluoride was almost entirely complexed with aluminum. As a result, levels of free fluoride were very low and decreased with decreasing pH. At higher solution pH values ($pH > 5.5$), fluoride ligands were unable to compete with hydroxide ligands for aluminum. In addition, aluminum hydrolysis resulted in a decrease in levels of inorganic monomeric aluminum. Consequently, free fluoride levels increased and aluminum-complexed fluoride levels decreased. Although levels of free and aluminum-complexed fluoride fluctuated considerably over the range of observed pH values, the sum of these two fluoride forms was relatively constant. The sum of free and aluminum-complexed fluoride approximates total fluoride. Although total fluoride measurements were made on only a few samples, for these samples, calculated and measured total fluoride values were similar. Silica, calcium, magnesium and iron complexes of fluoride are low in magnitude in Adirondack waters.

Table III. Fractional Composition of Aluminum Forms Based on Total Aluminum and Monomeric Aluminum Tabulated for Total Data, Lake Data and Stream Data

	Total Aluminum Data				*Lake Aluminum Data*				*Stream Aluminum Data*			
	Total Aluminum Based		*Monomeric Aluminum Based*		*Total Aluminum Based*		*Monomeric Aluminum Based*		*Total Aluminum Based*		*Monomeric Aluminum Based*	
Aluminum Form	*Mean*	*Std Dev*	*Mean*	*Std Dev*	*Mean*	*Std Dev*	*Mean*	*Std Dev*	*Mean*	*Std Dev*	*Mean*	*Std Dev*
Inorganic Monomeric	0.48	0.21	0.56	0.23	0.53	0.18	0.61	0.19	0.39	0.23	0.46	0.26
Organic Monomeric	0.37	0.19	0.44	0.23	0.33	0.17	0.39	0.19	0.45	0.21	0.54	0.26
Acid Soluble	0.14	0.12			0.14	0.13			0.16	0.10		
Free	0.10	0.08	0.11	0.09	0.10	0.08	0.11	0.09	0.08	0.08	0.10	0.09
Hydroxide Complexes	0.12	0.12	0.14	0.13	0.13	0.13	0.15	0.15	0.10	0.08	0.12	0.10
Fluoride Complexes	0.25	0.12	0.29	0.14	0.29	0.12	0.33	0.13	0.19	0.12	0.23	0.14
Sulfate Complexes	0.01	0.009	0.01	0.01	0.01	0.009	0.01	0.01	0.009	0.009	0.01	0.01

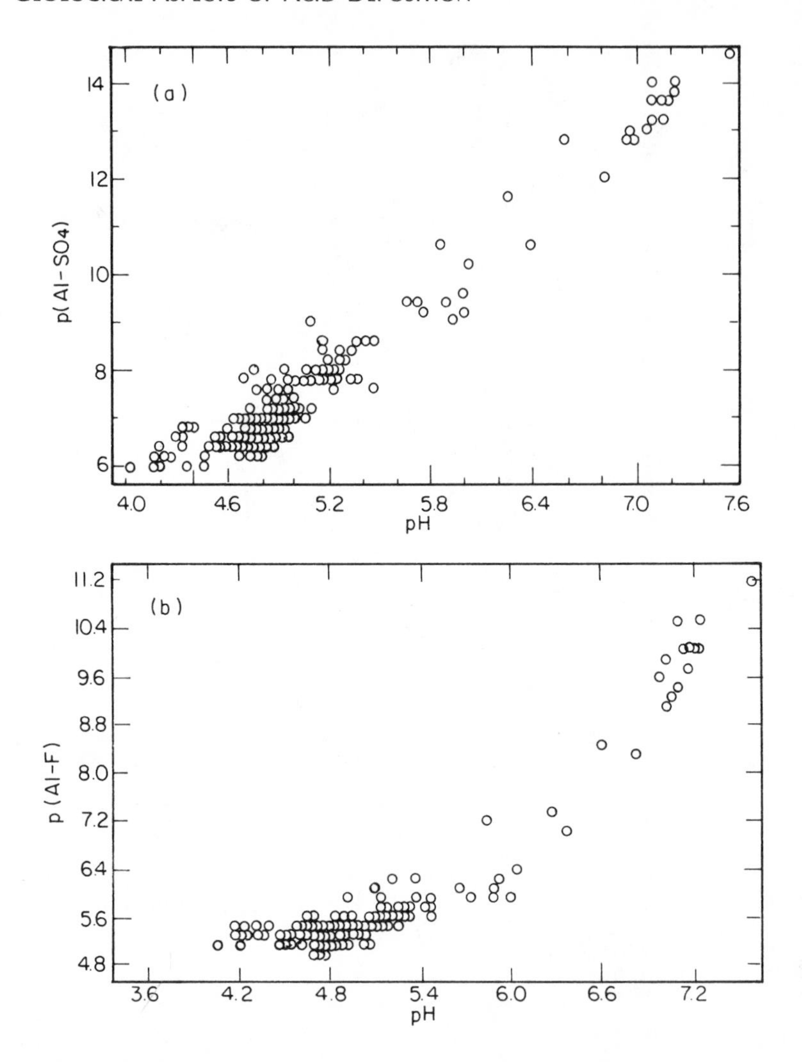

Figure 5. Sulfate ($Al\text{-}SO_4$) (a) and fluoride-complexed (Al-F) (b) aluminum levels in Adirondack surface waters as a function of solution pH.

Based on measured total concentrations of aluminum for the complete data set, the average charge per aluminum ion was +1.02 (acid soluble aluminum and aluminoorganic forms were assumed to carry no charge) (Table IV). Based on levels of monomeric aluminum, the average charge was +1.70. The mean charge per aluminum ion was pH-dependent, highly positive in low-pH solutions and decreasing in magnitude with increasing pH. Obviously, the assumption of a trivalent charge for a dissolved ion in dilute acidified waters is unrealistic. It is noteworthy that lakewater aluminum ions carried more charge than stream water aluminum.

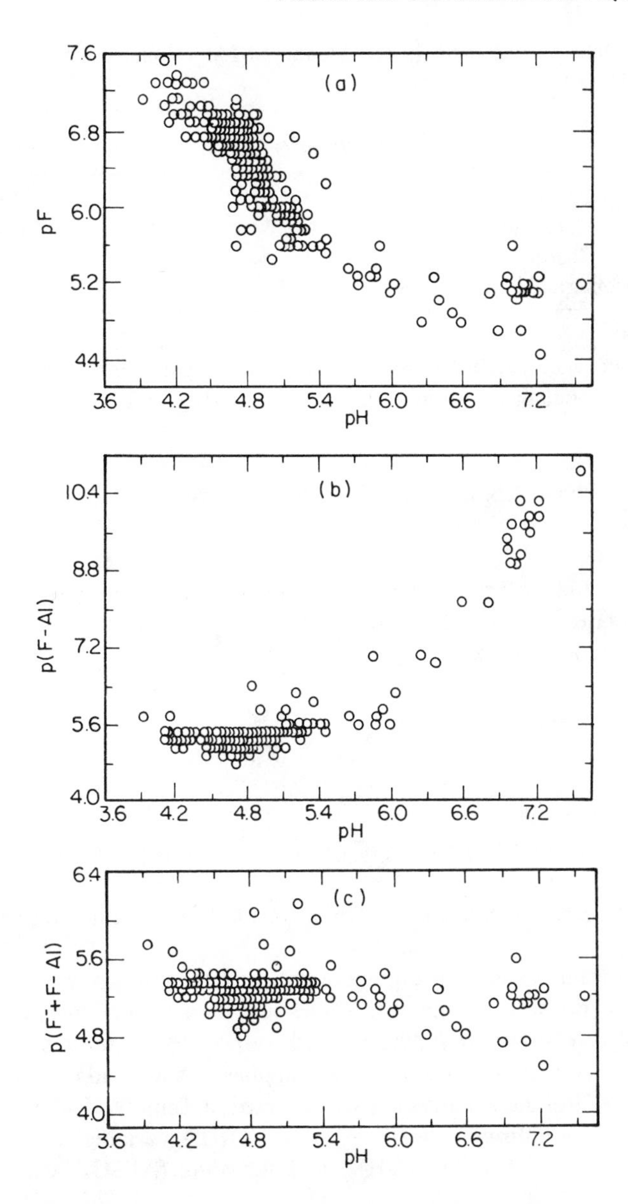

Figure 6. Free (F^-) (a), aluminum-complexed (F-Al) (b), and the sum of free and aluminum-complexed (c) fluoride levels in Adirondack surfaces waters as a function of solution pH.

Table IV. Average Charge Per Aluminum Ion Tabulated for Total Data, Lake Data and Stream Data

	Total Data	*Lake Data*	*Stream Data*
Average Charge per Total Aluminum Ion	1.02	1.10	0.87
Average Charge per Monomeric Aluminum Ion	1.70	1.75	1.61

This difference was attributed to the fact that stream water contained a higher percentage of noncharged aluminoorganic forms than did lakewater.

Mineral Phase Regulation of Aquoaluminum Levels

Disequilibrium indices are often used to compare water composition with respect to mineral phases and therefore postulate which mineral phase controls solution composition for a component of interest [28].

$$I_p = \log \frac{Q_p}{K_p}$$

where I_p = disequilibrium index
Q_p = ion activity product of the solution for a particular mineral phase of interest
K_p = thermodynamic equilibrium constant for the mineral phase of interest

A disequilibrium index greater than zero indicates that the system is oversaturated with respect to the specific mineral. Values less than zero denote undersaturation and values of zero suggest equilibrium with respect to the mineral of interest.

Disequilibrium index values for mineral phases that could regulate aquoaluminum levels in Adirondack waters are summarized in Table V. These mineral phases include $Al(OH)_3$, aluminum silicates ($Al_2Si_2O_5(OH)_4$), variable composition aluminosilicates, alunite ($KAl_3(SO_4)_2(OH)_6$) and jurbanite ($Al(SO_4)OH\ 5H_2O$) (Table II).

Adirondack surface waters were generally supersaturated with respect to kaolinite and undersaturated with respect to halloysite (Table V). Hem et al. [27] indicate that the stable form of $Al_2Si_2O_5(OH)_4$ is not readily synthesized at ambient temperature and pressure. Therefore, metastable forms like halloysite are more likely to be present in natural systems. If metastable forms regulate aquoaluminum levels, then $Al_2Si_2O_5(OH)_4$ would not appear to be a likely controlling phase. Dissolved silica levels (mean value 4.7 mg-L^{-1} of SiO_2) may be too low for the formation of halloysite. Hem et al. [27] suggest that if dissolved silica levels are

Table V. Mean Disequilibrium Index Values for Adirondack Surface Waters and a New Hampshire Stream (3) with Respect to Various Minerals[a]

	Adirondack Waters		
Mineral	*n*	*Ip*	*New Hampshire Stream Ip*
Amorphous Aluminum Trihydroxide	297	-2.31 ± 0.88	-2.14 ± 0.71
Microcrystalline Gibbsite	297	-0.86 ± 0.88	-0.69 ± 0.71
Natural Gibbsite	297	-0.28 ± 0.88	-0.11 ± 0.71
Synthetic Gibbsite	297	0.37 ± 0.88	0.55 ± 0.71
Halloysite	296	-1.30 ± 0.95	-1.08 ± 0.69
Kaolinite	296	1.04 ± 0.95	1.26 ± 0.69
Amorphous Alumino-silicate (ideal)	296	-1.33 ± 0.46	-1.24 ± 0.37
Reversible Alumino-silicate (non-ideal)	296	-0.69 ± 0.53	-0.59 ± 0.41
Jurbanite	297	-1.89 ± 2.07	-2.17 ± 1.69
Alunite	297	-1.48 ± 3.19	

[a] Ip values of zero indicate equilibrium, positive values indicate oversaturation, negative values indicate undersaturation. The error is expressed at the 95% confidence interval about the mean.

below 9.0 mg-L^{-1} of SiO_2, $Al_2Si_2O_5(OH)_4$ will not form. In addition, $Al_2Si_2O_5(OH)_4$ minerals were not observed in a limited X-ray diffraction examination of stream and lake sediments.

It is possible that other mineral phases regulate aquoaluminum levels (Table V). Basic aluminum sulfate ($Al(OH)SO_4$) has been reported to control aquoaluminum levels in acidic sulfate waters [30,34]. Eriksson [8] suggests that basic aluminum sulfate regulates aquoaluminum levels in acidified groundwaters in Sweden. Adirondack waters were greatly undersaturated with respect to aluminum sulfate mineral solubility (Table V). It is doubtful that alunite or jurbanite regulate aquoaluminum levels in Adirondack surface waters because the observed sulfate concentrations are low [30].

Paces [28] has proposed two variable composition aluminosilicate models: (1) an amorphous ideal aluminosilicate, and (2) a reversible nonideal aluminosilicate that may regulate aquoaluminum in natural waters. Adirondack waters were significantly undersaturated with respect to both these models; however, these phases exhibited the narrowest confidence intervals of any of the phases examined (Table V).

Aluminum trihydroxide may be important in regulating aquoaluminum levels in acidified waters. Aluminum levels in Adirondack surface waters were generally within the 95% confidence intervals for the solubility of synthetic gibbsite, natural gibbsite and microcrystalline gibbsite (Table V; Figure 7). Aluminum solubility appeared to be controlled by an aluminum trihydroxide phase slightly more soluble than synthetic gibbsite but less soluble than microcrystalline gibbsite.

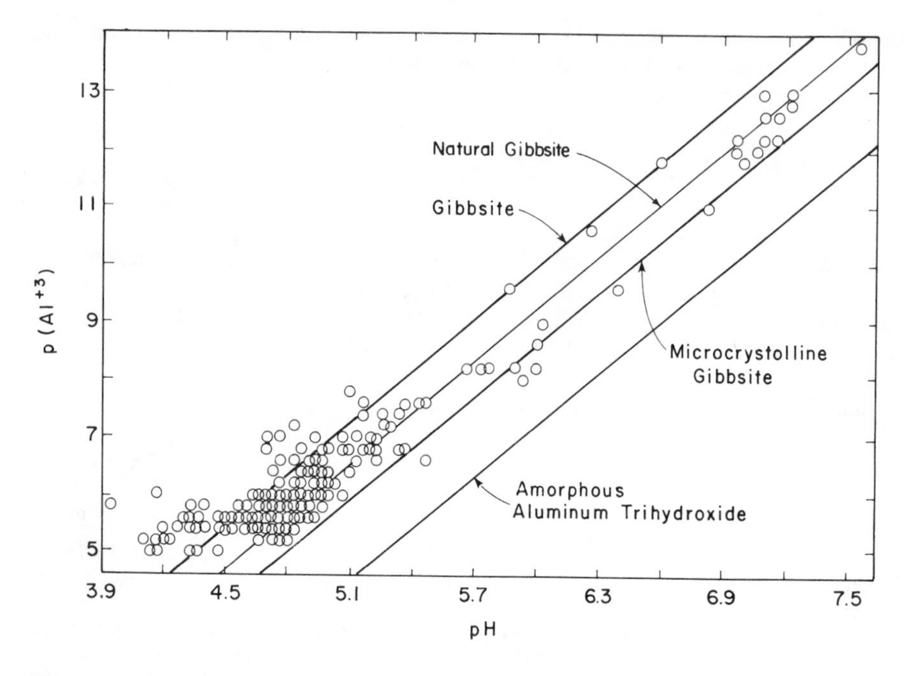

Figure 7. Water quality data of Adirondack surface waters superimposed on an $Al(OH)_3$ solubility diagram.

The aluminum solubility of a number of samples, however, fell outside the solubility band of gibbsite and microcrystalline gibbsite (Figure 7). Of all the samples, 15% were undersaturated with respect to the solubility of synthetic gibbsite, while only 2% of the samples were oversaturated with respect to the solubility of microcrystalline gibbsite.

The apparent solubility of aluminum trihydroxide ($3pH - pAl^{3+}$) decreased with decreasing pH values (Figure 8a). Many of the waters that were undersaturated with respect to synthetic gibbsite were low-pH samples. Low-pH samples were generally associated with rainfall and snowmelt events. It is hypothesized that samples collected at these times are less likely to be in equilibrium with soil minerals, due to insufficient time for equilibration [35]. Of the samples undersaturated with respect to aluminum trihydroxides (n = 46), 30% (n = 14) were surface lakewater samples collected in the winter and early spring and therefore diluted by snow (ice) meltwater. It is unlikely that these samples would be in equilibrium with any soil mineral phase.

The remainder of the samples undersaturated with respect to gibbsite were primarily samples from three streams [Little Creek (13 samples, 28%), West Pond Outlet (6 samples, 13%) and North Branch of the Black River (6 samples, 13%)]. It is noteworthy that two of these streams were high in organic carbon content (Little Creek, mean TOC = 9.3 mg-L^{-1} of C; and West Pond Outlet, mean TOC = 8.7 mg-L^{-1} of C). The data suggest that aluminum solubility ($3pH - pAl^{3+}$)

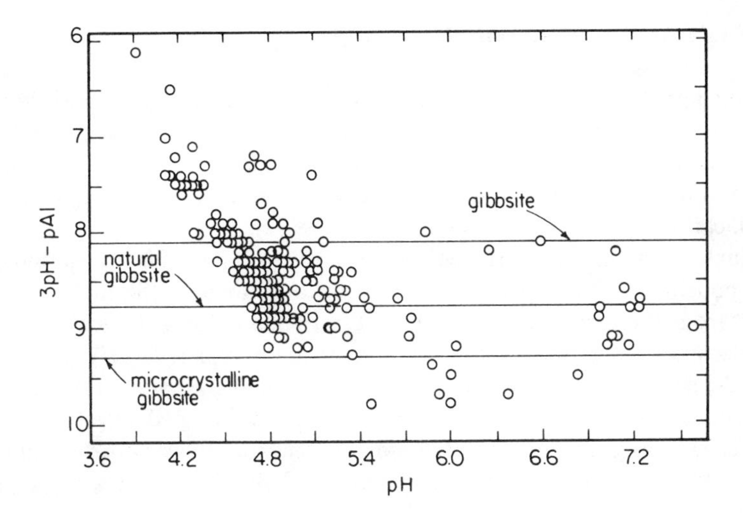

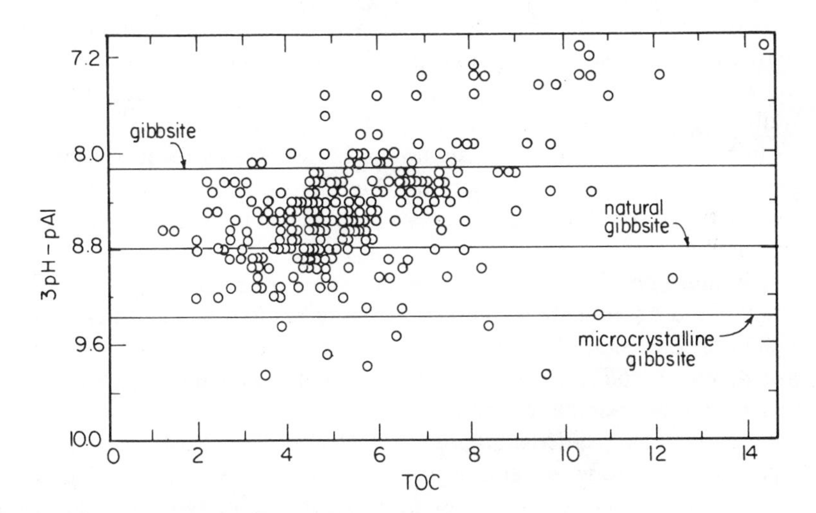

Figure 8. Aluminum solubility ($3pH - pAl^{3+}$) of Adirondack surface waters as a function of pH (a) and TOC concentration (b). Apparent aluminum solubility decreases with decreasing solution pH. Apparent aluminum solubility decreases with increasing organic carbon concentration lake meltwater samples, ($n = 14$) were excluded from this analysis ($3pH - pAl^{3+} = 9.05 - 0.10TOC$; $r^2 = 0.21$, $P < 0.0001$).

decreases with increasing organic carbon content: $3pH - pAl^{3+} = 9.05 - 0.10TOC$; $r^2 = 0.21$, $P < 0.0001$ (Figure 8b). This observation might reflect the flow paths of water through organic soil horizons, which are typically high in TOC and low in inorganic aluminum in the Adirondacks (Schofield, unpublished data).

DISCUSSION AND CONCLUSIONS

It has been hypothesized that mineral acids from atmospheric deposition have resulted in acidification of headwater lakes and streams in the northeastern United States [2,3,36]. A consequence of elevated aqueous hydrogen ion is increased mobilization of aluminum observed in Adirondack surface waters which may be a result of a modification of the soil podzolization process [2].

In high-elevation, northern temperature regions, the soils encountered are generally podzols [37]. The process of podzolization involves the mobilization of aluminum from the upper to the lower soil horizons by organic acids produced by litter decomposition. In lower soil horizons, aluminum is largely precipitated back into the soil [33,38]. A portion of organically complexed aluminum, however, may be leached from the soil and into surface waters during podzolization. During the summer months, when microbial decomposition activity was greatest and podzolization was intense, levels of TOC and associated aluminum increased in stream systems (Figure 4).

Superimposed on the natural podzolization process is the introduction of mineral acids to the edaphic environment through acidic precipitation or snowmelt. It is hypothesized that these acids mobilize very soluble aluminum previously precipitated within the soil during podzolization or on soil exchange sites. In dilute acidified waters, hydrogen ion concentration is highly variable, both spatially and temporally [3,35]. The aqueous chemistry of inorganic aluminum reflected these changes in pH. Levels of inorganic aluminum increased in streams during rainfall and snowmelt events, when waterflow was high and pH values were low. During low-flow, high-pH conditions in streams, inorganic monomeric aluminum levels were very low.

Aluminum solubility appeared to be regulated by an aluminum trihydroxide phase slightly more soluble than synthetic gibbsite but less soluble than microcrystalline gibbsite. Inclusion of impurities into the crystalline structure of aluminum trihydroxide may account for the increased solubility relative to synthetic gibbsite. As a result, natural minerals would be more soluble than synthetic minerals from which thermodynamic relationships are based. May et al. [22] compared the solubility of synthetic gibbsite with natural gibbsite (Minas Gerais, Brazil). The natural gibbsite was approximately four times more soluble than synthetic gibbsite and therefore fell between synthetic gibbsite and microcrystalline gibbsite with respect to solubility. In a laboratory study on the influence of mineral surfaces on aluminum chemistry, Brown and Hem [39] observed that the equilibria of synthetic aluminum solutions, in the presence of montmorillonite, volcanic ash, kaolinite and feldspathic sand, fell between the solubility range of microcrystalline gibbsite and synthetic gibbsite. These solutions were aged for 2–3 months.

Another possibility for the increased solubility of natural aluminum trihydroxide may be imperfect crystalline structure. If this were true it would help explain the lack of discernible gibbsite X-ray diffraction peaks in Adirondack lake and stream sediments.

In a recent field study of a small headwater stream in New Hampshire,

Johnson et al. [3] also found that an aluminum trihydroxide solubility model adequately described aquoaluminum levels. Ion activity product values observed by Johnson et al. [3] were similar to those posed by May et al. [22] for natural gibbsite. Disequilibrium indices for mineral phases were very similar in magnitude to those observed in our study (Table V). The mechanism by which aquoaluminum is regulated in headwater streams in New Hampshire may be similar to that in the Adirondacks.

The bulk of aluminum in Adirondack waters sampled was complexed with either natural organic matter or fluoride. These forms of aluminum are often ignored in studies of natural water as evidenced by several recent articles [2,7,8]. A study of aluminum equilibria in dilute acidified waters is seriously deficient unless organic and fluoride complexes are considered.

The elevation of inorganic aluminum levels in acidic Adirondack surface waters may have deleterious effects on fish indigenous to these ecosystems [12–14]. It appears that aluminum hydroxide complexes are more toxic to fish than are aluminum organic complexes or aluminum fluoride complexes [14]. Aluminum may also be detrimental to other aquatic biota inhabiting these waters [40]. In view of the biological repercussions of increased aqueous aluminum, an evaluation of mechanisms regulating solution concentration and a detailed analysis of aluminum speciation in natural waters are essential to our understanding of acidified ecosystems.

ACKNOWLEDGEMENT

Contribution No. 7 of the Upstate Freshwater Institute.

REFERENCES

1. Driscoll, C.T. "Aqueous Speciation of Aluminum in the Adirondack Region of New York State, USA," in *Proceedings of the International Conference on Ecological Impacts of Acid Precipitation,* D. Drablos and A. Tollan, Eds. (Oslo, Norway: SNSF Project, 1980), pp. 214–215.
2. Cronan, C.S., and C.L. Schofield. "Aluminum Leaching Response to Acid Precipitation: Effects on High Elevation Watersheds in the Northeast," *Science* 204:304–306 (1979).
3. Johnson, N.M., C.T. Driscoll, J.S. Eaton, G.E. Likens and W.H. McDowell. "Acid Rain, Dissolved Aluminum and Chemical Weathering at the Hubbard Brook Experimental Forest, New Hampshire," *Geochim. Cosmochim. Acta* 45:1421–1437 (1981).
4. Seip, H.M. "Acidification of Freshwater—Sources and Mechanisms," in *Proceedings of the International Conference on Ecological Impacts of Acid Precipitation,* D. Drablos and A. Tollan, Eds. (Oslo, Norway: SNSF Project, 1980), pp. 358–366.
5. Hultberg, H. and S. Johansson. "Acid Groundwater," *Nordic Hydrol.* 12:51–64 (1981).
6. Robertson, C.E., and J.D. Hem. "Solubility of Aluminum in the Presence of Hydroxide, Fluoride, and Sulfate," U.S. Geol. Surv. Water Supply Paper 1827-C (1969).

7. Johnson, N.M. "Acid Rain: Neutralization Within the Hubbard Brook Ecosystem and Regional Implications," *Science* 204:497–499 (1979).
8. Eriksson, E. "Aluminum in Groundwater: Possible Solution Equilibria," *Nordic Hydrol.* 12:43–50 (1981).
9. Driscoll, C.T. "Chemical Characterization of Some Dilute Acidified Lakes and Streams in the Adirondack Region of New York State," PhD Thesis, Cornell University (1980).
10. Johannessen, M. "Aluminum, A Buffer in Acidic Waters," in *Proceedings of the International Conference on Ecological Impacts of Acid Precipitation,* D. Drablos and A. Tollan, Eds. (Oslo, Norway: SNSF Project, 1980), pp. 222–223.
11. Dickson, W. "Some Effects of the Acidification of Swedish Lakes," *Verh. Int. Verein. Limnol.* 20:851–856 (1978).
12. Schofield, C.L., and J.R. Trojnar. "Aluminum Toxicity to Fish in Acidified Waters," in *Polluted Rain,* T.Y. Toribara, M.W. Miller and P.E. Morrow, Eds. (New York: Plenum Press, 1980), pp. 347–366.
13. Baker, J.P., and C.L. Schofield. "Aluminum Toxicity to Fish as Related to Acid Precipitation and Adirondack Surface Water Quality," in *Proceedings of the International Conference on Ecological Impacts of Acid Precipitation,* D. Drablos and A. Tollan, Eds. (Oslo, Norway: SNSF Project, 1980), pp. 292–293.
14. Driscoll, C.T., J.P. Baker, J.J. Bisogni and C.L. Schofield. "Effect of Aluminum Speciation on Fish in Dilute Acidified Waters," *Nature* 284:161–164 (1980).
15. Isachsen, Y.M., and D.W. Fisher. "Geologic Map of New York—Adirondack Sheet," Map and Chart Series No. 15, New York State Museum of Science (1970).
16. Gran, G. "Determination of the Equivalence Point in Potentiometric Titrations," *Int. Cong. Anal. Chem.* 77:661–671 (1952).
17. "Orion Instruction Manual, Fluoride Electrodes," Orion Research Inc., Cambridge, MA (1976).
18. *Standard Methods for the Examination of Water and Wastewater,* 14th ed. (New York: American Public Health Association, 1976).
19. Kahn, H.L. "Principles and Practice of Atomic Absorption," in *Trace Inorganics in Water,* R.A. Baker, Ed., Advances in Chemistry Series 106 (Washington, DC: American Chemical Society, 1968), pp. 183–229.
20. Rainwater, F.H., and L.L. Thatcher. "Methods of Collection and Analysis of Water Samples," U.S. Geol. Surv. Water Supply Paper 1454 (1960), pp. 97–100.
21. Smith, R.W. "Relations Among Equilibrium and Non-equilibrium Aqueous Species of Aluminum Hydroxide Complexes," in *Nonequilibrium Systems in Natural Water Chemistry,* J.D. Hem, Ed., Advances in Chemistry Series 106 (Washington, DC: American Chemical Society, 1971), pp. 250–279.
22. May, H.M., P.A. Helmke and M.L. Jackson. "Gibbsite Solubility and Thermodynamic Properties of Hydroxy-Aluminum Ions in Aqueous Solution at 25°C," *Geochim. Cosmochim. Acta* 43:861–868 (1979).
23. Hem, J.D. "Graphical Methods for Studies of Aqueous Aluminum Hydroxide, Fluoride and Sulfate Complexes," U.S. Geol. Surv. Water Supply Paper 1827-B (1968).
24. Behr, B., and H. Wendt. "Fast Ion Reactions in Solution (I) Formation of the Aluminum Sulfate Complexes," *Zeits. Elekrochem.* 66:223–228 (1962).
25. Hem, J.D., and C.E. Robertson. "Form and Stability of Aluminum Hydroxide Complexes in Dilute Solution," U.S. Geol. Surv. Water Supply Paper 1827-A (1967).
26. Stumm, W., and J.J. Morgan. *Aquatic Chemistry* (New York: John Wiley & Sons Inc., 1970).
27. Hem, J.D., C.E. Robertson, C.J. Lind and W.L. Polzer. "Chemical Interactions of

Aluminum with Aqueous Silica at 25°C," U.S. Geol. Surv. Water Supply Paper 1827-E (1973).
28. Paces, T. "Reversible Control of Aqueous Aluminum and Silica During the Irreversible Evolution of Natural Waters," *Geochim. Cosmochim. Acta* 42:1487–1493 (1978).
29. Adams, F., and Z. Rawajfih. "Basaluminite and Alunite: A Possible Cause of Sulfate Retention by Acid Soils," *Soil Sci. Soc. Am. J.* 41:686–692 (1977).
30. Nordstrom, D.K. "The Effect of Sulfate on Aluminum Concentrations in Natural Waters: Some Stability Relations in the System Al_2O_3–SO_3–H_2O at 298°K," *Geochim. Cosmochim. Acta* 46:681–692 (1982).
31. Livingston, D.A. "Chemical Composition of Rivers and Lakes," U.S. Geol. Surv. Professional Paper 440–6 (1963).
32. Lind, C.J. U.S. Geol. Surv. Professional Paper 700D (1970), pp. 272–280.
33. Lind, C.J., and J.D. Hem. "Effects of Organic Solutes on Chemical Reactions of Aluminum," U.S. Geol. Surv. Water Supply Paper 1827-G (1975).
34. Yan Breemen, N. "Dissolved Aluminum in Acid Sulfate Soils and in Acid Mine Waters," *Soil Sci. Soc. Am. Proc.* 37:694–697 (1973).
35. Johnson, N.M., G.E. Likens, F.H. Bormann, D.W. Fisher and R.S. Pierce. "A Working Model for the Variation in Stream Water Chemistry at the Hubbard Brook Experimental Forest, New Hampshire," *Water Resources Res.* 5:1353–1363 (1969).
36. Schofield, C.L. "Acid Precipitation: Our Understanding of the Ecological Effects," in *Proceedings of a Conference on Emerging Environmental Problems: Acid Precipitation,* EPA-902/9-75-001 (1975), pp. 76–87.
37. Buckman, H.O., and N.C. Brady. *The Nature and Properties of Soils* (New York: Macmillan, 1961).
38. Ugolini, F.C., R. Minden, H. Dawson and J. Zachara. "An Example of Soil Processes in the *Abies Amabilis* Zone of Central Cascades, Washington," *Soil Sci.* 724:291–302 (1977).
39. Brown, D.W., and J.D. Hem. "Reactions of Aqueous Aluminum Species at Mineral Surfaces," U.S. Geol. Surv. Water Supply Paper 1827-F (1975).
40. Hall, R.J., C.T. Driscoll and G.E. Likens. "Physical, Chemical and Biological Consequences of Episodic Aluminum Additions to a Stream Ecosystem," *Limnol. Oceanog.* (submitted).

CHAPTER 5

Ion Balances Between Precipitation Inputs and Rhode River Watershed Discharges

David L. Correll
Nancy M. Goff
William T. Peterjohn

In recent years, environmental scientists in several locations have gradually perceived the importance of chemical fluxes that enter various ecosystems in precipitation. To a considerable extent this concern has resulted from the documentation of steadily increasing acidity in rainfall [1]. This increased flux of hydrogen ions in precipitation is primarily due to increasing concentrations of sulfur and nitrogen oxides in the atmosphere [2]. In such places as Sweden [3,4], the White Mountains of New Hampshire [5–7], the Smoky Mountains of North Carolina [8,9], the Appalachian Mountains of Tennessee [9], the Rocky Mountains in Colorado [10] and the Sangre de Cristo Mountains of New Mexico [11] research reports have documented and summarized both ionic inputs in precipitation and ionic losses in land discharge. Differences between ionic inputs and outputs can then be ascribed to the interactions of vegetation and soils with chemical components in the precipitation. Most of these published studies were conducted in mountainous regions with low human populations and limited land management. None were in the Atlantic Coastal Plain of the United States. An understanding of natural (i.e., unmanaged) systems is theoretically important but of limited value when extrapolated to complex, multiple-land-use systems. The land area of the United States is a mosaic of primarily three land uses: pasture/rangeland (39%), ungrazed forest (21%) and cropland (17%) [12]. Thus understanding the nutrient dynamics of multiple-land-use basins is critically important for wise management of the land.

In the early 1970s the Smithsonian Institution initiated long-term environmental studies of the Rhode River Ecosystem in Maryland with an emphasis on systems analysis (Figure 1). This system is located on the coastal plain and the most abundant land uses on the watershed are forest, cropland and pastureland. The forest is composed of mixed deciduous, broadleaved species typical of the eastern U.S.

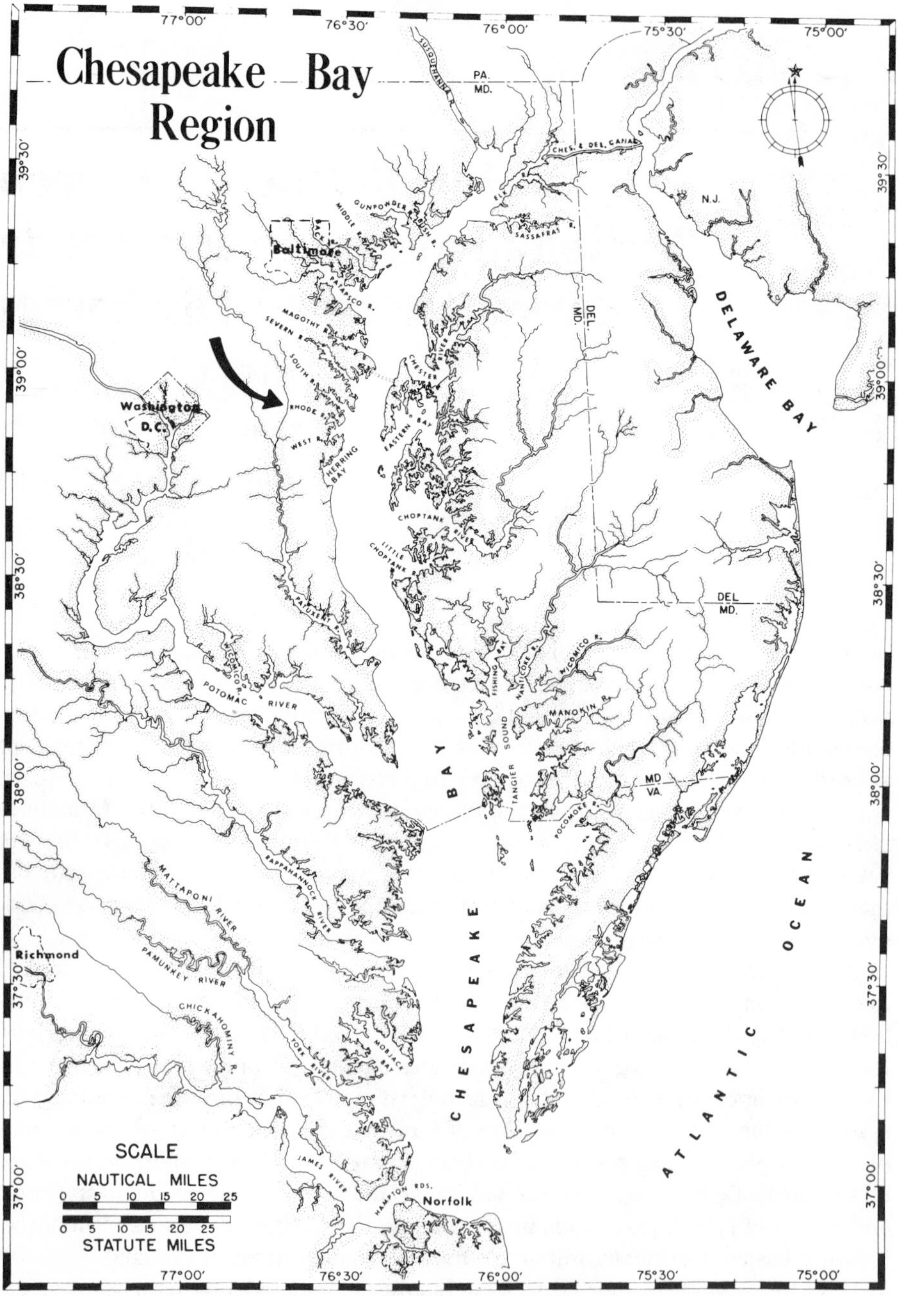

Figure 1. Map of the Chesapeake Bay Region. Location of Rhode River Ecosystem is marked with an arrow.

deciduous forest. Since no calcareous minerals are found in the watershed soils, they are poorly buffered against acid rain inputs. Past research on the Rhode River system has included the effects of nutrients in precipitation on the estuary [13] and the watershed [14,15]. It also has documented changes in the pH of precipitation and stream discharges for 7–10 years.

The objectives of the project reported here were:

1. a comparison of ion inputs from precipitation with outputs in land discharge for a mature forest;
2. assessing how cropland and pasturelands differ in their ion balances due to management practices; and
3. tracing changes in ionic composition as water moves down through the watershed either over the soil/litter surface during storms or through the soils as shallow groundwater.

Available data indicate that the coastal plain region might be especially sensitive to displacement of potassium, magnesium and calcium by hydrogen ions, since these elements are present only in low concentrations in the soils and no calcium-containing minerals occur in the surface soils. Thus, it was anticipated that the data resulting from this study would allow us to test whether important losses of these essential plant nutrients were occurring from the watershed.

In addition to the possible effects of cation displacements, another concern involves the increasing flux of nitrate and sulfate from the atmosphere. Nitrate is a significant source of nitrogen to forests and both sulfate and nitrate can serve as electron acceptors in anoxic soils.

A long-term objective is to understand the impact of management practices such as liming and fertilizing not only on chemical ion balances, but also on plant and animal species that are adapted to a specific set of ion-balance and pH conditions. The ability to predict changes at the ionic level could result in more efficient management and a more stable ecosystem.

SITE DESCRIPTION

The Rhode River Estuary is a tidal tributary to Chesapeake Bay (Figure 1), located approximately 20 km south of Annapolis, MD (38°53′ N, 76°35′ W). Geologically, the Rhode River watershed is located on the inner mid-Atlantic Coastal Plain. It has sedimentary soils from the Pleistocene Talbot formation at low elevations on the eastern portion of the watershed, Eocene Nanjemoy formation soils at low elevations further west, Miocene Calvert formation soils at intermediate elevations and Pleistocene Sunderland formation soils at the highest elevations. A few outcrops of Pleistocene Wicomico formation soils are also found. The mineralogy of the surface soils in the watershed is fairly uniform with high levels of montmorillonite [$Al_2Si_4O_{10}(OH)_2$], intermediate levels of illite (interlayered mica and montmorillonite) and kaolinite ($H_4Al_2Si_2O_9$), and lower levels of plagioclase

($NaAlSi_3O_8$), potassium feldspar ($KAlSi_3O_8$), gibbsite [$Al(OH)_3$] and chlorite [$(Al_2Si_6)Al_2Mg_{10}O_{20}(OH)_{16}$] in the silt and clay fractions. The soils differ locally primarily with respect to the proportions present of sand, silt and clay. Underlying the watershed is an impervious clay layer (the Marlboro Clay) which acts as an effective aquaclude [16]. A layer of glauconitic sands [$K_{15}(Fe,Mg,Al)_{4-6}(Si,Al)_8 O_{20}(OH)_4$] is found immediately above the Marlboro Clay.

Three small subwatersheds representative of the most abundant land uses on the watershed and in the Atlantic Coastal Plain were selected for study. All three are within the Eocene Nanjemoy formation and have fine sandy loam soils, which have been intensively mapped by the Soil Conservation Service. One study site (watershed) was composed of 6.3 ha of mixed, broadleaved deciduous forest. An area on the upper part (0.6 ha) was farmed until approximately 1940, then abandoned. The remaining 5.7 ha were never clear-cut or farmed [17,18]. The average watershed slope is 8.3%. The mean surface soil pH of the forest site was 4.9 and surface organic matter content was 4.2%. The soils of the lower elevations are from the Keyport series, those at intermediate elevations are from the Howell and Donlonton series, while those at the highest elevations are from the Monmouth, Adelphia and Collington series.

The second study watershed (6.1 ha) was composed of 4.4 ha of perennial cattle pasture and 1.7 ha of grazed riparian forest. Surface soil pH was 5.3 and surface soil organic matter content was 3.5%. Average watershed slope was 10.8%. The soils of the lower elevations are from the Monmouth series, those of the intermediate elevations are from the Howell series and the upper elevation soils are from the Marr and Westphalia series.

The third study watershed (16.3 ha) was composed of 10.4 ha of cornfields under conventional tillage and 5.9 ha of hedgerows and riparian forest. Surface soil pH was 5.6 and surface soil organic matter content was 1.9%. The soils of the lower part are primarily of the Collington series while those of the upper part of the watershed are primarily of the Westphalia series. Average watershed slope is 5.4%.

SAMPLING

Bulk precipitation was sampled continuously at an elevation of 13 m as described previously [13]. Land discharge from the three study watersheds was monitored and sampled at permanent 120° V-notch weirs. The weirs included stilling wells and instrument sheds equipped with depth monitors (Stevens model 7001) which recorded digital data on paper tape every 5 min, and flowmeters (Stevens model 61R) that had been modified to close a sampling switch once every 38,000 L. Each switch closure activated a sampling cycle in which a fixed volume of stream water was pumped from the bottom of the V-notch into sample bottles. Samples were composited for one-week intervals over a one-year study period. One bottle contained sulfuric acid preservative for analysis of biologically labile parameters [19] and another had no preservatives. At times of very low discharge weekly spot samples were taken. If stream stage height was above a predetermined threshold, a custom-built fraction collector was also activated with each flowmeter switch closure. In this case up to 12 discrete bottles without preservative could be taken automatically at flow-spaced intervals. Each switch closure during which the fraction collector was active caused an event mark to be made on the paper punch tape.

Transects of surface water collectors and shallow groundwater wells were established on each study watershed. Each transect consisted of several clusters of samplers located from high to low ground along the expected direction of flow. Thus, uphill/downhill comparisons could be made as well as interwatershed comparisons. Each cluster consisted of three replicate wells and three surface collectors, approximately 10 m from each other along a line normal to the transect axis. Surface water collectors were 4-L polyethylene bottles placed into holes in the soil in an inverted position. A slot was cut at ground level and an apron of plastic sheeting sealed to the bottom of the slot was spread uphill a short distance to funnel surface runoff into the bottle. Plastic tubing was sealed into the mouth of the inverted bottle so that samples could be withdrawn without disturbing the samplers. Before withdrawing samples, air was briefly bubbled through the sample to suspend particulates immediately before the sample was withdrawn. Samplers were rinsed out between storm events. Groundwater wells (piezometers) were used to sample shallow groundwater. They consisted of 3.8-cm-i.d. polyvinyl chloride (PVC) pipe perforated with 2.5-mm holes for approximately 8 cm on the lower end. The bottom end of the pipe was capped. Holes were bored with a bucket auger either to the top of the Marlboro Clay or to 6 m depth, whichever came first. The pipes were then inserted and clay was packed around the pipe at the soil surface. The pipe was then cut off approximately 0.4 m above the surface and a loose cap was placed over the top. In most cases another well, which was shallower in depth, was placed in a cluster. Our intent was to test realistically whether lateral, and perhaps vertical variances in soil water and surface runoff water composition were less than differences found between clusters within a given transect or between different transects. Two transects were established on the cropland watershed, one with three clusters of samplers and one with two clusters. Only one transect was used on each of the other two study watersheds, each consisting of only two clusters. Wells were pumped out one day before sample collection.

ANALYSIS OF SAMPLES

Na^+, K^+, Ca^{2+}, Mg^{2+}, NO_3^-, NO_2^-, Cl^- and SO_4^{2-} were analyzed with a Dionex Model 16 ion chromatograph after filtration through prewashed Millipore® HA membrane filters (0.45 mm nominal pore size). Nitrite was usually only present in trace amounts. Therefore, nitrate and nitrite were routinely summed and will be referred to as nitrate. Millipore HA-filtered water was also analyzed for total Kjeldahl nitrogen by digestion as described by Martin [20], distillation and Nesslerization. Ammonia was determined by oxidation to nitrite and subsequent coupling to sulfanilamide [21]. Total phosphorus was determined by perchloric acid digestion and reaction with ammonium molybdate and stannous chloride [22]. Orthophosphate was determined by the same procedure without the digestion and organic phosphorus was calculated as total P minus orthophosphate. Hydrogen ion was determined with an expanded-range pH meter and bicarbonate by titration. Only about 10% of the samples were analyzed for bicarbonate. The rest were estimated from pH data with a regression derived from samples that were analyzed (meq HCO_3^- equals $-0.80 + 0.163$pH. This regression had an R^2 of 0.82. Approximately 10% of compositional data for other ion parameters were estimated. These were estimated as seasonal averages or by interpolation in most cases. When flowrates were changing rapidly, a few values were estimated as being the same as found at other times under similar flow conditions.

RESULTS

pH of Precipitation and Land Discharge

An essentially continuous data record for bulk precipitation pH at the Rhode River site exists for the last seven years. The mean pH declined over that time period for each season of the year, but the trend is most apparent for the spring season when precipitation results primarily from regional fronts (Figure 2). Yearly variation in pH was apparent and high variance within each season was also found. Mean seasonal pH declined at a rate of 0.12 pH unit per year and the correlation had a R^2 of 0.92. A seven-year record from a slightly earlier period is also available for stream water pH in the streams draining the larger watersheds of the Rhode River site. Between 1972 and 1978, weekly data from five to seven streams were taken and mean seasonal pH was calculated. Dry years such as 1977 had higher mean pH, and summer and fall data were erratic. However, in the winter and spring, when significant runoff occurred, stream pH values seem to have declined about 0.5 units during those years (Figure 3). If one examines the winter and spring stream pH data for the last five years from the mature, forested watershed from which ion balance data were taken in this study (Figure 4), a decline is also apparent. Although perhaps due to chance, the rates of decline in the pH of this forested stream are almost identical to that of bulk precipitation (Figure 2). This stream was not studied earlier.

These pH data illustrate that (1) the Rhode River site is well within the so-called acid rain belt of eastern North America; (2) these coastal plain soils are not able to buffer effectively the acidity of the precipitation; and (3) therefore, the mid-Atlantic Coastal Plain is an area very vulnerable to acid rain.

Hydrologic Data

Summaries of fluxes into the watershed study sites as bulk precipitation and out of the sites as combined surface and groundwater drainage are presented in Table

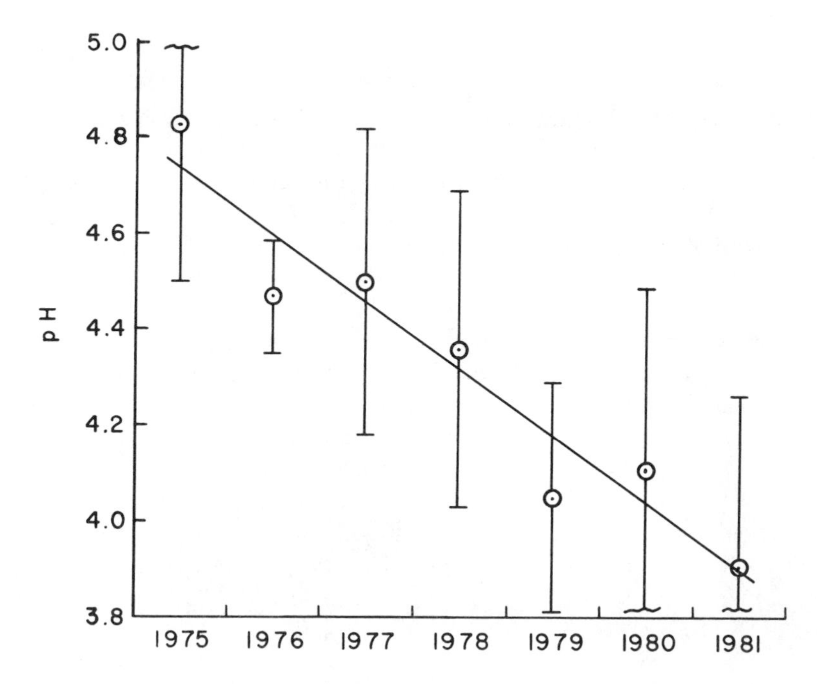

Figure 2. Mean pH of bulk precipitation in the spring at the Rhode River site. Bars are one standard deviation. $R^2 = 0.92$.

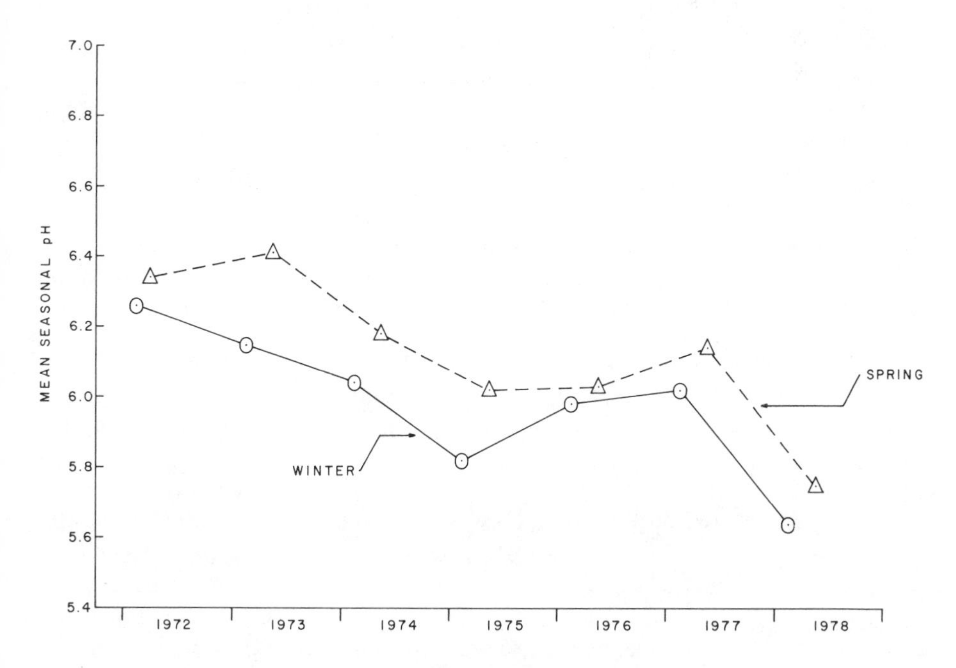

Figure 3. Mean seasonal pH of larger streams draining the Rhode River watershed.

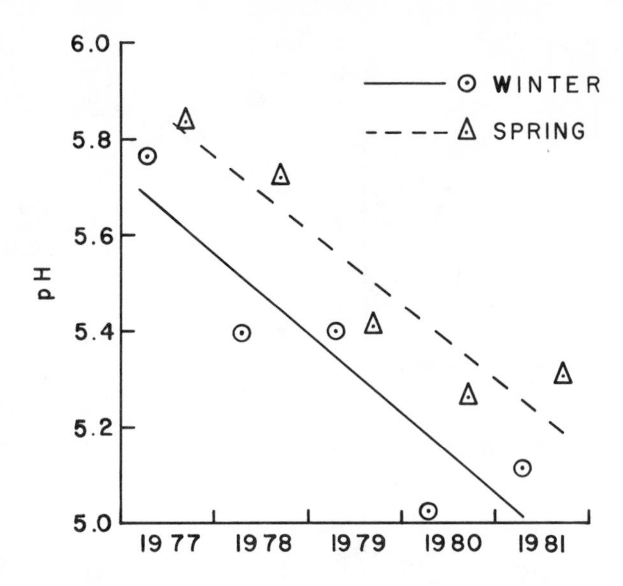

Figure 4. Mean seasonal pH of stream draining mature forest study site watershed. Correlation for spring, −0.94; for winter, −0.91.

I. Precipitation volume during the study period (100.4 cm) was somewhat below the 160-year long-term mean of 108 cm [23]. The winter season had slightly above average precipitation, while all others, especially fall, were below average. During the preceding year, only 87.8 cm of precipitation occurred and during the season immediately before the study period (winter of 1980–1981) only 12.6 cm of precipitation occurred. Thus, following a year of drought, soils were dry and water tables were low when this study was initiated and until late in the study period land discharges were low.

Ion Balances of Mature Forest

Although cation input fluxes were highest in spring (0.85 $keq\text{-}ha^{-1}$), and next highest in the summer (0.48 $keq\text{-}ha^{-1}$), cation outputs were highest in winter and spring. Less than 15% of the measured cation inputs were discharged during the study period. This low discharge was probably largely due to the dry conditions and should be reversed during the first extended period of wet weather. No discharges at all occurred during the fall, but with increased rains during the winter months discharges steadily increased. Of the annual cation outputs, 40% were during February alone. Hydrogen ions comprised 50% of cation inputs but less than 1% of cation outputs. Organic nitrogen (20%), ammonium (10%), calcium and sodium (6% each), magnesium (5%), and potassium (2%) constituted the remainder of the measured cation inputs.

Table I. Summary of Ion Fluxes (keq-ha^{-1}) in Bulk Precipitation and Land Discharges from Three Study Watersheds Between March 1, 1981, and February 28, 1982.

Season	*Vol. (cm)*	Na^+	K^+	*Organic* N^a	NH_4^+	H^+	Mg^{2+}	Ca^{2+}	*Sum Cations*	NO_3^-	$H_2PO_4^-$	*Organic* P^a	HCO_3^-	Cl^-	SO_4^{2-}	*Sum Anions*
Bulk Precipitation																
Spring	25.4	0.073	0.029	0.23	0.083	0.38	0.030	0.034	0.85	0.10	0.00019	0.0069	0.00	0.11	0.20	0.42
Summer	29.7	0.027	0.00088	0.086	0.051	0.26	0.020	0.028	0.48	0.11	0.0009	0.0008	0.000	0.099	0.21	0.42
Fall	18.8	0.012	0.0017	0.036	0.036	0.16	0.014	0.017	0.28	0.081	0.0011	0.0006	0.000	0.056	0.21	0.35
Winter	26.5	0.019	0.0027	0.026	0.036	0.14	0.021	0.026	0.27	0.15	0.0008	0.0004	0.000	0.055	0.23	0.44
Year	100.4	0.12	0.042	0.38	0.21	0.94	0.085	0.11	1.88	0.45	0.0046	0.0086	0.000	0.32	0.86	1.64
Mature Forest Land Discharge																
Spring	1.46	0.038	0.013	0.0043	0.0008	0.0009	0.031	0.020	0.11	0.0014	0.0001	0.0003	0.0027	0.033	0.085	0.12
Summer	1.44	0.014	0.010	0.0050	0.0013	0.0004	0.014	0.011	0.056	0.0021	0.0017	0.0002	0.0042	0.012	0.038	0.0580
Fall	0.00															
Winter	1.86	0.020	0.0090	0.0044	0.0009	0.0009	0.050	0.032	0.12	0.0044	0.0001	0.0001	0.0052	0.035	0.14	0.18
Year	4.76	0.072	0.033	0.014	0.0030	0.0021	0.095	0.063	0.28	0.0080	0.0019	0.0006	0.0122	0.080	0.27	0.37
Pastureland Discharge																
Spring	3.20	0.064	0.030	0.0073	0.0014	0.0016	0.034	0.037	0.18	0.012	0.0001	0.0002	0.0081	0.058	0.17	0.25
Summer	2.72	0.030	0.015	0.0054	0.0015	0.0004	0.022	0.030	0.10	0.0054	0.0002	0.0003	0.0235	0.025	0.066	0.12
Fall	0.91	0.0074	0.0050	0.0013	0.0005	0.0001	0.013	0.017	0.044	0.0009	0.0001	0.0001	0.0115	0.0093	0.034	0.056
Winter	4.46	0.027	0.010	0.0046	0.0009	0.0020	0.049	0.050	0.14	0.035	0.0001	0.0002	0.0109	0.044	0.30	0.39
Year	11.3	0.13	0.061	0.018	0.0042	0.0041	0.12	0.13	0.47	0.053	0.0004	0.0007	0.054	0.14	0.57	0.82
Cropland Discharge																
Spring	6.93	0.15	0.083	0.041	0.014	0.0012	0.12	0.14	0.55	0.048	0.0036	0.0035	0.051	0.29	0.33	0.72
Summer	4.01	0.074	0.025	0.023	0.0063	0.0010	0.11	0.11	0.35	0.055	0.0009	0.0009	0.024	0.19	0.18	0.45
Fall	0.66	0.0091	0.0051	0.0008	0.0002	0.0002	0.024	0.019	0.058	0.015	0.0000	0.0000	0.0014	0.039	0.058	0.10
Winter	11.6	0.11	0.037	0.022	0.0044	0.0043	0.36	0.29	0.82	0.18	0.0003	0.0003	0.017	0.57	0.78	1.54
Year	23.2	0.35	0.15	0.088	0.024	0.0067	0.61	0.56	1.78	0.28	0.0048	0.0048	0.093	1.08	1.35	2.82

[a] It was assumed that all organic N or P was monovalent.

In forest outputs, magnesium was the most abundant cation (34%), followed by sodium (26%), calcium (22%), potassium (12%), organic N (5%) and ammonium (1%). If chloride can be assumed to be reasonably inert chemically and biologically, the ratio of a given cation to chloride can be used as a tracer for enrichment or depletion due to watershed interactions. The Mg^{2+}/Cl^- ratio was enriched the most over precipitation (4.5-fold) followed by potassium (3.1-fold), then calcium and sodium (both 2.4-fold). The ratios of organic N and ammonium ion to chloride in discharge were less than in precipitation. Ratios of potassium and magnesium to chloride were enriched the most over precipitation in the summer (9.6-fold and 5.8-fold, respectively). The highest potassium-to-chloride ratios were observed during a major summer storm event when most of the flow was due to surface runoff (Table II). At that time the K^+/Cl^- ratio was enriched 20-fold and the Mg^{2+}/Cl^- ratio was enriched 10-fold over the monthly average for precipitation. Surface and shallow groundwater flowrates were calculated from the hydrograph by the method of Barnes [24] for all storm events during the study period. A linear regression of K^+/Cl^- ratios in weekly spring and summer forest discharges vs the ratio of surface to groundwater discharge had a correlation of 0.71 and gave the equation $K^+/Cl^- = 0.30 + 0.467$(surface/groundwater ratio).

In general, both the potassium-to-sodium and calcium-to-magnesium ratios in forest runoff were higher in summer. Perhaps this was due to the larger proportion (57%) of land discharge which was surface runoff. The K^+/Na^+ ratio varied more than the Ca^{2+}/Mg^{2+} ratio. Both the K^+/Cl^- and Na^+/Cl^- ratios declined in forest discharge in winter. The decline in K^+/Cl^- was greater, which may indicate a greater role for biota in the release of potassium than sodium.

Anion inputs in bulk precipitation were dominated by sulfate (52%) and nitrate (27%), followed by chloride (20%) and contained less than 1% of combined inorganic and organic phosphates (Table I). Input fluxes were relatively constant from season to season.

Due to drought conditions, anion outputs from the forest study site (Table I) were only 22% of inputs. Sulfate constituted 75% of these outputs and chloride 22%, while nitrate was only 2%. Chloride outputs only exceeded inputs in February, the last month of the study period, when 1.5 cm of discharge occurred and salts stored in the soils due to evapotranspiration during the drought began to be flushed out. Sulfate-to-chloride ratios varied relatively little from season to season or with flowrates during storms and were only enriched over precipitation by 25% for the year.

Ion Balances of Pastureland

Cation discharges from the pasture study site watershed were somewhat larger than from the forest site (Table I), but the volume of water discharged was also greater, due to lower rates of evapotranspiration. During the study period no synthetic fertilizer or lime was applied and the cattle were not fed concentrates.

Table II. Land Discharge Data on Sequential Samples from Two Storms: Mature Forest Watershed

Fraction	*Time*	*Discharge (L-s^{-1})*	*Flow Ratio, Quick/Slow*	*Na^+ (meq-L^{-1})*	*Na^+/Cl^-*	*K^+ (meq-L^{-1})*	*K^+/Cl^-*	*K^+/Na^+*	*Mg^{2+} (meq-L^{-1})*	*Mg^{2+}/Cl^-*	*Ca^{2+} (meq-L^{-1})*	*Ca^{2+}/Cl*	*Ca^{2+}/Mg^{2+}*
May 15, 1981, Storm (4.98 cm, peak intensity of 1.55 mm-min^{-1})													
1	1500	26	8.8	0.078	0.61	0.066	0.52	0.85	0.132	1.04	0.089	0.70	0.67
2	1530	16	5.2	0.077	0.61	0.074	0.58	0.96	0.154	1.21	0.099	0.78	0.64
3	1830	2.0	0.0	0.139	1.04	0.063	0.47	0.45	0.276	2.06	0.158	1.18	0.57
July 4, 1981, Storm (10.06 cm, peak intensity of 1.30 mm-min^{-1})													
1	0935	34	4.6	0.057	1.46	0.065	1.67	1.14	0.057	1.46	0.058	1.49	1.02
2	0950	54	6.7	0.077	1.79	0.077	1.79	1.00	0.074	1.72	0.068	1.58	0.92
3	1040	56	7.5	0.065	1.51	0.074	1.72	1.14	0.085	1.98	0.069	1.60	0.81
4	1100	58	8.2	0.090	1.48	0.077	1.26	0.86	0.085	1.39	0.069	1.13	0.81
5	1115	22	2.9	0.080	1.43	0.080	1.43	1.00	0.097	1.73	0.075	1.34	0.77
6	1700	2.8	0.0	0.117	1.36	0.078	0.91	0.67	0.131	1.52	0.100	1.16	0.76

However, during prior years these types of inputs sometimes occurred and may influence some of the data. Between 22 and 26 cattle were pastured on an 18.9-ha pasture, which included the study watershed site, for the entire study period, except during April through July when no cattle were present. For the entire year of the study an average of 0.91 adult cow equivalents per hectare were pastured. Cation outputs of sodium and calcium (28% each) were highest and magnesium outputs (26%) were almost as high. Potassium (13%), organic nitrogen (4%) and ammonium (1%) constituted the remaining cation discharges measured. These values are rather similar to those for forest discharges, as were the ratios of K^+, Na^+, Ca^{2+} and Mg^{2+} to Cl^-. On the average, however, Mg^{2+}/Cl^- ratios were somewhat lower and Ca^{2+}/Cl^- ratios were somewhat higher than for forest discharges. Calcium-to-magnesium ratios in pasture discharges averaged 1.13 for the year with little seasonal difference, while the ratio averaged 0.66 for forest discharges. Compositional data for discharges from two storm events is given in Table III. Potassium-to-chloride ratios peaked in discharges during storm events with a maximum enrichment over summer bulk precipitation of 38-fold during the July storm event. The May storm is a clear example of the changes in K^+ and Na^+ concentrations during a storm event and of how they relate to surface flow and the groundwater (slow flow) recession curve (Figure 5). Generally, Na^+ concentrations seemed to change more than K^+ during storms. Sometimes K^+ concentrations didn't change much at all. Before the May storm (Figure 5), both K^+ and Na^+ concentrations were high, then the concentrations both declined during peak surface flow, but Na^+ declined much more than K^+. Then, during the hydrograph recession, K^+ concentration fell further while Na^+ began to increase. A few days later, when groundwater flowrates were very low again, Na^+ continued to increase and K^+ began to increase. The K^+/Na^+ ratio was always highly correlated with discharge rate and during the May storm high K^+/Na^+ ratios were found in discharges at the peak of the storm when most was due to surface runoff. The ratio declined steadily with the declining hydrograph. The highest Mg^{2+}/Cl^- ratios were found on the descending part of the hydrograph for the May storm, corresponding to a 4.6-fold enrichment over spring bulk precipitation. Unfortunately, the sampler failed during part of the July storm, so that the falling hydrograph was not analyzed chemically. The spring average K^+/Na^+ ratio for pasture discharges was 0.47 which is only 20% above that for spring bulk precipitation. Both K^+/Na^+ and Ca^{2+}/Mg^{2+} ratios averaged higher in the summer and fall for discharge from pastureland than during the winter and spring. As in the case of forest discharges, this winter decline in K^+/Na^+ ratio was due to a greater decline in the K^+/Cl^- than the Na^+/Cl^- ratio.

Anion outputs from pastureland (Table I) were only 47% of precipitation inputs. Over half of the measured outputs occurred in the winter when the volume of discharges began to increase. As for forest, sulfate constituted 75% and chloride 18% of anion discharges. Nitrate, however, was more significant (7%). Sulfate-to-chloride ratios were highest in winter and lowest in summer. For the year, sulfate-to-chloride in discharges was enriched 55% over precipitation.

Table III. Land Discharge Data on Sequential Samples from Two Storms: Pastureland Watershed

Fraction	*Time*	*Discharge* $(L\text{-}s^{-1})$	*Flow Ratio, Quick/Slow*	Na^+ $(meq\text{-}L^{-1})$	Na^+/Cl^-	K^+ $(meq\text{-}L^{-1})$	K^+/Cl^-	K^+/Na^+	Mg^{2+} $(meq\text{-}L^{-1})$	Mg^{2+}/Cl^-	Ca^{2+} $(meq\text{-}L^{-1})$	Ca^{2+}/Cl^-	Ca^{2+}/Mg^{2+}
May 15, 1981, Storm													
1	1500	17	19	0.027	0.31	0.069	0.79	2.56	0.056	0.64	0.055	0.63	0.98
2	1515	16	15	0.026	0.09	0.068	0.25	2.62	0.081	0.30	0.085	0.31	1.05
3	1530	3.0	2.3	0.077	0.66	0.064	0.55	0.83					
4	1615	0.90	0	0.105	0.81	0.053	0.41	0.50	0.134	1.04	0.127	0.98	0.95
5	1730	0.70	0	0.101	0.83	0.042	0.35	0.42	0.147	1.21	0.158	1.31	1.07
6	1930	0.60	0	0.098	0.93	0.038	0.36	0.39	0.134	1.28	0.143	1.36	1.07
7	2400	0.45	0	0.105	0.89	0.035	0.30	0.33	0.149	1.26	0.153	1.30	1.03
July 4, 1981, Storm													
1	0830	0.75	0	0.070	1.84	0.113	2.97	1.61	0.041	1.08	0.063	1.66	1.54
2	0950	13	4.9	0.039	1.95	0.040	2.00	1.03	0.022	1.10	0.031	1.55	1.41
3	1005	13	4.9	0.069	2.88	0.083	3.46	1.20	0.022	0.92	0.024	1.00	1.09
4	1020	4.3	1.1	0.080	2.58	0.091	2.94	1.14	0.033	1.06	0.042	1.35	1.27
5	1035	8.6	3.4	0.063	2.10	0.081	2.70	1.29	0.026	0.87	0.035	1.17	1.35

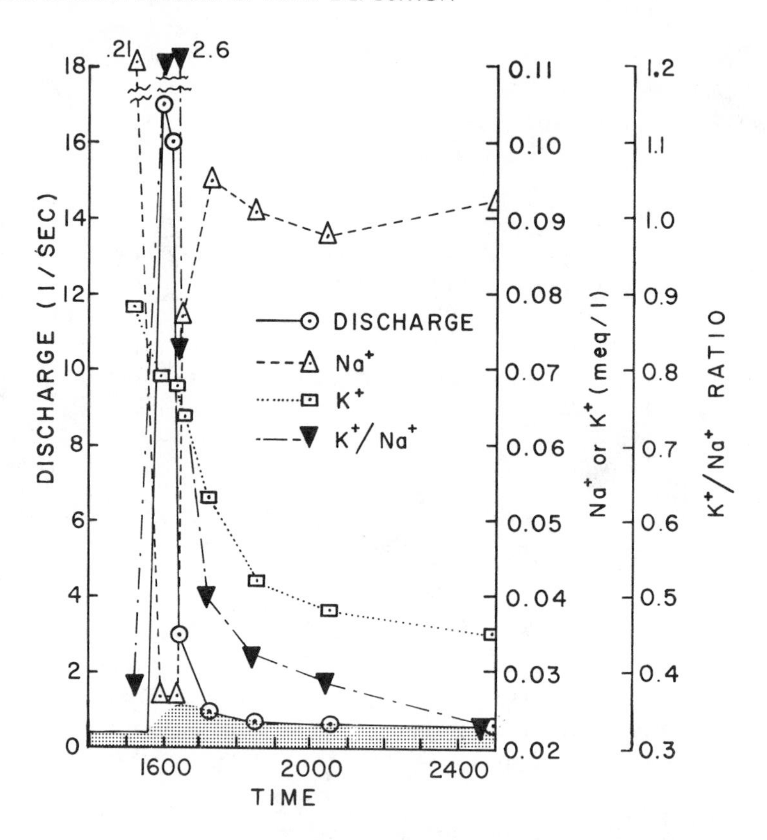

Figure 5. Changes in surface runoff (quick flow) and groundwater (slow flow) discharge rates and concentrations of K^+ and Na^+ in the stream draining the pastureland study site occurring during the May 15, 1981, storm event. Groundwater discharge is shaded and surface runoff discharge is the difference between the total discharge and the shaded area.

Ion Balances of Cropland

Cation discharges from the cropland study site (Table I) were much higher than from the other two sites. The largest cation discharges were in the winter (46% of the total for the year) and 28% of the total cation discharges occurred in February. Almost five times more volume of water than the forest site was discharged, helping to explain the greater ion yields. Higher ion discharges were also partially due to the management input loadings. Those occurred in April and May for all but the organic N, which was applied in July. Management loadings were estimated to be (1) 19 keq-ha^{-1} of Ca^{2+} and 2.7 keq-ha^{-1} of Mg^{2+} from applications of limestone and (2) 0.15 keq-ha^{-1} of K^+, 3.3 keq-ha^{-1} of organic N, 2.2 keq-ha^{-1} of ammonium, 1.1 keq-ha^{-1} of NO_3^- and 0.40 keq-ha^{-1} of $H_2PO_4^-$ in farm fertilizer. Analysis of granular fertilizer indicated the additional presence of significant amounts of other ions. Approximately 0.45 keq-ha^{-1} of Cl^- and 1.0 keq-ha^{-1} of SO_4^{2-} were added in granular fertilizers. Other agricultural chemicals may also have contained significant amounts of chloride, since more chloride (1.08 keq-ha^{-1}) was measured in discharges than in precipitation plus fertilizer (0.77 keq-ha^{-1}). Cation outputs were greatest for magnesium (34%) followed by calcium (31%), sodium (20%), potassium (80%), organic N (5%), ammonium (1%) and traces of H^+. Ratios of cations to chloride were not a very useful tracer for enrichment, since a large amount of chloride inputs was in the form of agricultural chemicals, which are applied at different times and in differing ratios to other ions of interest. Compared with cation inputs from precipitation and management, very little ammonium, organic N or hydrogen ion was discharged. The average ratio of calcium to magnesium in cropland discharges (0.92) was intermediate between that from forest (0.66) and pastureland (1.08); however, the seasonal pattern differed being highest in the spring (1.2) and lowest in winter, reaching 0.66 in December. The highest values observed (1.8–1.9) were in samples from the second peak in the July storm event (Table IV), but even these were well below the ratio of 7.0 applied as limestone. Ratios of K^+/Na^+ were similar to those found for discharges from pastureland (Table I). Management inputs of potassium were in mid-April and K^+ discharge was highest in spring, but the K^+/Na^+ ratio was highest in the fall and at times of high storm discharge (Table IV, Figures 6 and 7). The K^+/Na^+ ratio was lowest in June (0.28) and February (0.32).

Anion outputs from cropland were composed of sulfate (48%), chloride (38%), nitrate (10%), bicarbonate (3%) and traces (0.3%) of phosphates. Of the anion discharge, 55% occurred during winter. Despite large management inputs, sulfate was a lower proportion of total cropland anion discharges than for forest or pastureland.

Intrawatershed Ion Patterns

Forest

Data on the ionic composition of groundwater in the upper part of the watershed and both surface runoff and groundwater in the lower part of the forested watershed are summarized in Table V. At the lower elevation all cations except sodium were in higher concentrations in groundwater and of these the largest was a 12-fold increase in ammonium ion. Hydrogen ion and bicarbonate data were not tabulated since they were not routinely determined for these samples. However, five sets of pH measurements of groundwater taken from August 1981 to February 1982 show an average increase from 5.1 at the upper position to 6.2 at the lower position. Total measured groundwater anions decreased going down the elevational gradient. Of these nitrate and sulfate concentrations declined 7- and 5-fold, respectively, whereas chloride and phosphate concentrations increased 1.4- and 38-fold, respectively. Groundwater ratios of K^+/Na^+ and Ca^{2+}/Mg^{2+} increased along the elevational gradient primarily because of a concomitant decline in the Na^+/Cl^-

Table IV. Land Discharge Data on Sequential Samples: Cropland Watershed (July 4, 1981, Storm Only)

Fraction	*Time*	*Discharge* ($L\text{-}s^{-1}$)	*Flow Ratio, Quick/Slow*	Na^+ ($meq\text{-}L^{-1}$)	Na^+/Cl^-	K^+ ($meq\text{-}L^{-1}$)	K^+/Cl^-	K^+/Na^+	Mg^{2+} ($meq\text{-}L^{-1}$)	Mg^{2+}/Cl^-	Ca^{2+} ($meq\text{-}L^{-1}$)	Ca^{2+}/Cl^-	Ca^{2+}/Mg^{2+}
1	0650	62	0	0.074	0.52	0.077	0.54	1.04	0.105	0.74	0.142	1.00	1.35
2	0740	80	0	0.096	0.60	0.076	0.48	0.79	0.123	0.77	0.160	1.00	1.30
3	0850	46	0	0.116	0.50	0.074	0.32	0.64	0.149	0.64	0.167	0.72	1.12
4	0930	130	0.34	0.062	0.65	0.058	0.61	0.94	0.085	0.89	0.122	1.28	1.44
5	0945	440	1.8	0.053	0.96	0.056	1.02	1.06	0.057	1.04	0.084	1.53	1.47
6	0950	540	2.2	0.054	1.06	0.066	1.29	1.22	0.071	1.39	0.095	1.86	1.34
7	1000	530	1.7	0.041	0.75	0.061	1.11	1.49	0.052	0.95	0.078	1.42	1.50
8	1010	450	1.5	0.039	0.65	0.066	1.10	1.69	0.053	0.88	0.069	1.15	1.30
9	1030	120	0.21	0.066	0.85	0.074	0.95	1.12	0.058	0.74	0.080	1.03	1.38
10	1100	160	0.37	0.077	0.73	0.071	0.68	0.92	0.077	0.73	0.141	1.34	1.83
11	1115	240	0.51	0.043	0.61	0.060	0.86	1.40	0.073	1.04	0.138	1.97	1.89
12	1140	170	0.26	0.059	0.83	0.075	1.06	1.27	0.067	0.94	0.085	1.20	1.27

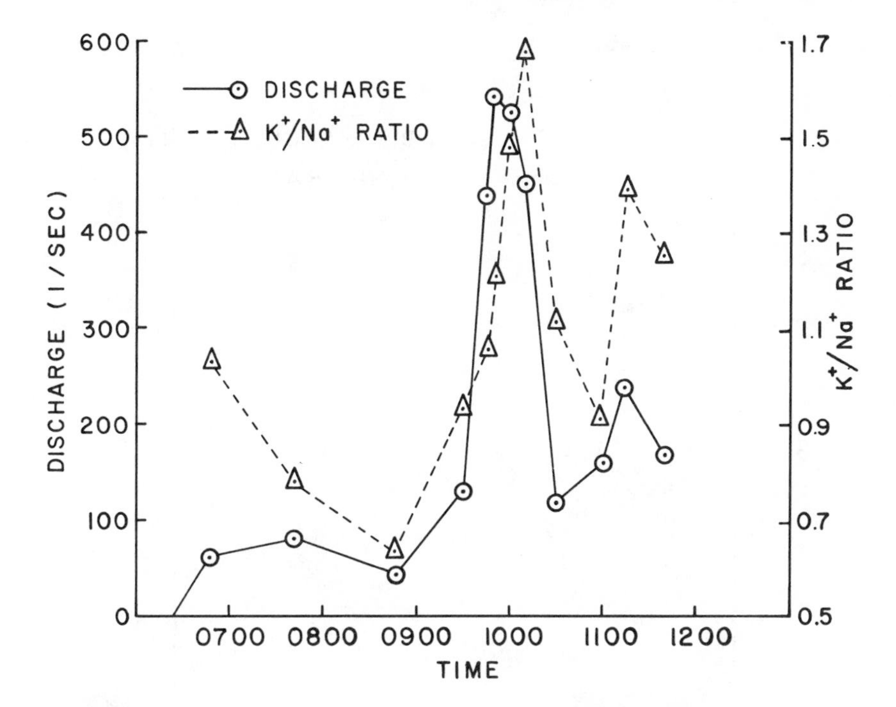

Figure 6. Changes in the hydrograph and K^+/Na^+ ratios for the stream draining the cropland site during the July 4, 1981, storm event, subsequent to being dry.

ratio and increase in Ca^+/Cl^- ratio. Changes in groundwater ion concentrations along the hydrologic gradient appear to be controlled by the combined effects of pH and redox potential. Low redox potentials would account for the lowered concentrations of NO_3^- (via denitrification) and SO_4^{2-} (via reduction to H_2S) and also the increased concentration of NH_4^+ (via a shutdown in nitrification). Higher pH values could explain the higher concentrations of $H_2PO_4^-$ due to reduced binding by Fe, Mn and Al compounds in the deeper subsoils.

Surface runoff contained lower concentrations of sodium and higher concentrations of potassium, ammonium, organic nitrogen and nitrate than the groundwater at the lower elevation. The K^+/Na^+ ratio in surface runoff averaged 3.45 during the study year and 5.36 during the spring. Respectively, these ratios were 9.8 and 22 times higher than found in groundwater, and 9.9 and 14 times higher than found in bulk precipitation. Concentrations of nitrate in surface runoff were lower and sulfate higher than those measured in bulk precipitation.

Pastureland

The most noticeable differences in groundwater data between the pasture study site (Table VI) and the forest site are the somewhat lower concentration of cations

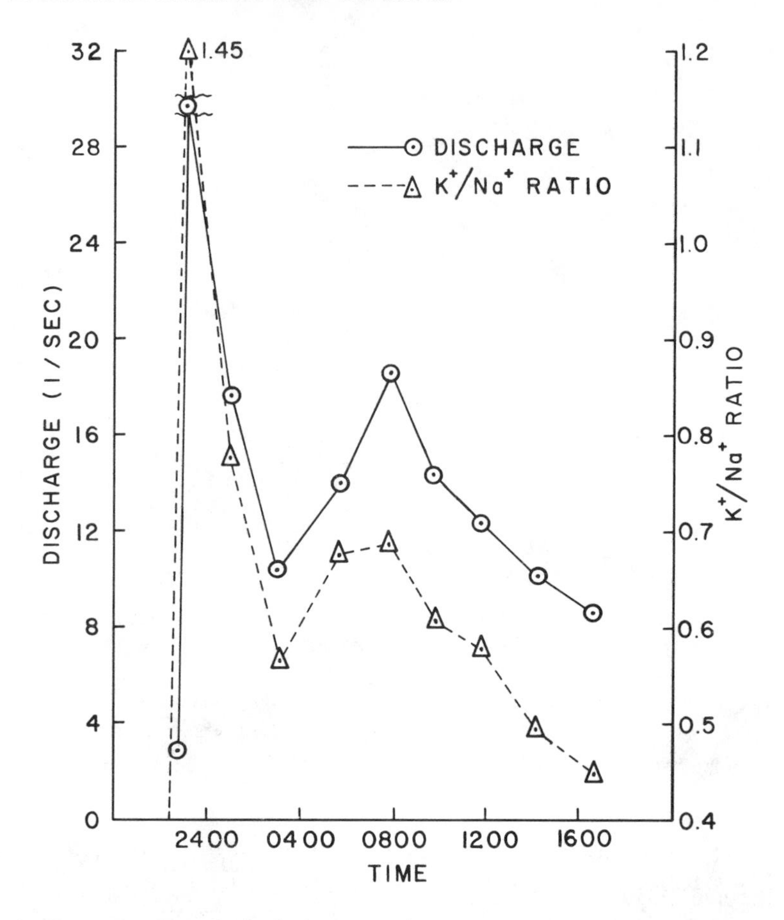

Figure 7. Changes in the hydrograph and K^+/Na^+ ratios on the stream draining the cropland study site during the February 2–3, 1982, storm event.

measured at the lower elevations and the reversal of the gradient patterns found for phosphate, magnesium, calcium and chloride concentrations.

Groundwater at the lower elevation had 3.2-fold higher ammonium concentrations and 3.8-, 4.5-, 2.6-, 1.2-, 1.3- and 1.2-fold lower concentrations, respectively for phosphate, calcium, magnesium, chloride, sodium and sulfate. Five sets of groundwater samples for the higher position and three for the lower also showed a decline in average pH from 6.3 to 5.5. Less anoxic conditions in groundwater at the lower elevation probably account for the smaller decline in SO_4^{2-} and increase in NH_4^+ concentrations when compared to the forest. The decline of pH with elevation in the pasture could also explain the reversed pattern of phosphate concentrations.

Surface runoff for the year contained higher concentrations of all cations and anions, except for NO_3^- and SO_4^{2-} when compared to the groundwater at the

Table V. Summary of Ionic Composition ($meq\text{-}L^{-1}$) of Surface Runoff and Shallow Groundwaters: Mature Forest Watershed

		Na^+	K^+	Organic N	NH_4^+	Mg^{2+}	Ca^{2+}	Sum of Cations	NO_3^-	Total P[a]	Cl^-	SO_4^{2-}	Sum of Anions
Watershed Elevation of 15 m													
Groundwater	Spring	0.240	0.050	0.008	0.003	0.213	0.141	0.655	0.010	0.000	0.123	0.580	0.714
	Summer	0.240	0.043	0.010	0.004	0.112	0.098	0.507	0.016	0.000	0.164	0.606	0.786
	Fall	0.166	0.037	0.002	0.001	0.110	0.091	0.408	0.015	0.002	0.108	0.612	0.738
	Winter	0.122	0.028	0.003	0.001	0.131	0.096	0.381	0.014	0.000	0.128	0.736	0.878
	Year	0.192	0.039	0.006	0.002	0.141	0.107	0.488	0.014	0.001	0.131	0.636	0.781
Watershed Elevation of 5 m													
Surface Runoff	Spring	0.061	0.327	0.114	0.003	0.122	0.135	0.762	0.008	0.007	0.322	0.099	0.436
	Summer	0.103	0.275	0.219	0.061	0.143	0.177	0.978	0.043	0.020	0.117	0.186	0.365
	Fall	0.069	0.199	0.131	0.146	0.194	0.296	1.035	0.026	0.050	0.239	0.205	0.520
	Year	0.077	0.266	0.155	0.070	0.153	0.202	0.924	0.026	0.026	0.226	0.163	0.440
Groundwater	Spring	0.256	0.063	0.020	0.023	0.232	0.298	0.893	0.002	0.020	0.165	0.169	0.355
	Summer	0.215	0.081	0.023	0.017	0.169	0.277	0.782	0.000	0.024	0.209	0.098	0.330
	Fall	0.160	0.071	0.008	0.031	0.205	0.264	0.739	0.000	0.067	0.191	0.100	0.358
	Winter	0.116	0.049	0.008	0.026	0.244	0.317	0.760	0.004	0.043	0.171	0.115	0.339
	Year	0.187	0.066	0.015	0.024	0.212	0.289	0.793	0.002	0.038	0.186	0.120	0.346

[a] Most phosphorus was orthophosphate.

Table VI. Summary of Ionic Composition (meq-L^{-1}) of Surface Runoff and Shallow Groundwaters: Pastureland Watershed

		Na^+	K^+	Organic N	NH_4^+	Mg^{2+}	Ca^{2+}	Sum of Cations	NO_3^-	Total P[a]	Cl^-	SO_4^{2-}	Sum of Anions
Watershed Elevation of 22 m													
Groundwater	Summer	0.189	0.052	0.021	0.004	0.133	0.336	0.736	0.033	0.022	0.108	0.425	0.589
	Fall	0.142	0.043	0.014	0.007	0.264	0.486	0.956	0.010	0.045	0.159	0.472	0.686
	Winter	0.129	0.040	0.013	0.007	0.287	0.539	1.01	0.024	0.046	0.117	0.548	0.735
	Year	0.153	0.045	0.016	0.006	0.228	0.454	0.902	0.022	0.038	0.128	0.482	0.704
Watershed Elevation of 7.5 m													
Surface Runoff	Spring	0.452	0.033	0.102	0.034	0.217	0.100	0.937	0.018	0.004	0.936	0.192	1.15
	Summer	0.199	0.309	0.137	0.023	0.132	0.200	1.00	0.037	0.015	0.087	0.182	0.321
	Fall	0.048	0.481	0.151	0.014	0.166	0.392	1.25	0.003	0.035	0.128	0.120	0.286
	Year	0.233	0.274	0.130	0.023	0.172	0.231	1.06	0.019	0.018	0.384	0.165	0.586
Groundwater	Spring	0.192	0.075	0.026	0.032	0.081	0.074	0.479	0.100	0.001	0.118	0.311	0.440
	Summer	0.131	0.052	0.028	0.030	0.067	0.098	0.406	0.024	0.001	0.093	0.373	0.491
	Fall	0.072	0.035	0.005	0.013	0.089	0.103	0.316	0.038	0.001	0.089	0.458	0.585
	Winter	0.072	0.026	0.008	0.003	0.111	0.123	0.343	0.022	0.000	0.110	0.532	0.665
	Year	0.116	0.047	0.017	0.019	0.087	0.100	0.386	0.023	0.001	0.103	0.418	0.545

[a] Most phosphorus was orthophosphate.

same elevation. Nitrate concentration was only 82% and sulfate 39% as high as in groundwater. Seasonal variations in the concentrations of surface runoff were particularly high for Na^+, K^+ and Cl^-, each having more than a 0.4-meq range in concentration. The dominant cations and anions were sodium and chloride during the spring, potassium and sulfate during the summer, and potassium and chloride during the fall. Compared to the forest, ion concentrations in pastureland runoff were very similar differing substantially only in sodium (three times higher) and ammonium (threefold decrease). Potassium-to-chloride and potassium-to-sodium ratios were greater than one in the surface runoff and less than one in the groundwater for all seasons except spring. In spring, surface runoff contained a very low concentration of potassium resulting in K^+/Na^+ and K^+/Cl^- ratios of less than one and less than the groundwater ratios.

Cropland

Data from the two transects on the cropland site were averaged and are presented in Table VII. Groundwater concentrations underwent little seasonal variation at either elevation. The only seasonal pattern was a steady, twofold decline in sodium concentrations from spring to winter, which was a pattern common to all watersheds in this study. Groundwater leaving the cultivated portion of the cropland each season contained high calcium, magnesium, sulfate and nitrate concentrations. After traversing 50–100 m of riparian forest, nitrate, magnesium and calcium concentrations were reduced 94, 33 and 37%, respectively, whereas ammonium ion increased fivefold. Five sets of groundwater samples also showed an average increase in pH from 4.8 to 5.3 between the bottom of the cropland and the stream bank. The large reductions in nitrate and smaller concomitant increases in ammonium and pH indicate that low redox potentials and high levels of denitrification probably occurred in the riparian zone soils.

Surface runoff leaving the fields had higher concentrations of K^+, organic N, NH_4^+ and total P than groundwater had leaving the cultivated fields. Surface water that had traversed the riparian forest had higher concentrations of K^+, organic N, NO_3^- and total P than did groundwater at the same location. Transit through the riparian forest caused average yearly reductions in all ion concentrations, except total P. However, substantial declines (greater than a 2-fold reduction) occurred only for the nitrate (6-fold) and ammonium (6.4-fold) ions. Seasonal variations in concentrations were often greater than twofold at both elevations, and the direction and magnitude of the riparian forest's effect on the water flowing through it also varied. Compared to surface runoff concentrations in the forest and pasture study sites, surface water leaving the cropland contained substantially higher levels of Mg^{2+}, NO_3^-, and SO_4^{2-} and slightly higher levels of Ca^{2+} and Cl^-. After passing through the riparian zone, surface runoff concentrations were lower than those in the forested watershed for all ions except NO_3^- and SO_4^{2-}. As was typical in the other two watersheds K^+/Cl^- and K^+/Na^+ ratios were higher for surface runoff than for groundwater.

Table VII. Summary of Ionic Composition (meq-L^{-1}) of Surface Runoff and Shallow Groundwaters: Cropland Watershed

		Na^+	K^+	*Organic N*	NH_4^+	Mg^{2+}	Ca^{2+}	*Sum of Cations*	NO_3^-	*Total P*[a]	Cl^-	SO_4^{2-}	*Sum of Anions*
At Bottom of Cropland													
Surface Runoff	Spring	0.047	0.151	0.105	0.259	0.134	0.200	0.896	0.266	0.008	0.223	0.332	0.829
	Summer	0.067	0.204	0.194	0.084	0.358	0.454	1.36	0.749	0.004	0.197	0.569	1.52
	Fall	0.044	0.183	0.056	0.059	0.131	0.157	0.630	0.114	0.004	0.327	0.282	0.725
	Year	0.053	0.180	0.118	0.134	0.207	0.270	0.962	0.376	0.005	0.249	0.394	1.02
Groundwater	Spring	0.192	0.076	0.019	0.006	0.467	0.352	1.11	0.400	0.005	0.334	0.459	1.20
	Summer	0.172	0.069	0.010	0.007	0.343	0.273	0.874	0.470	0.002	0.364	0.418	1.25
	Fall	0.138	0.060	0.010	0.006	0.344	0.274	0.832	0.504	0.006	0.376	0.556	1.44
	Winter	0.092	0.030	0.011	0.002	0.441	0.338	0.914	0.646	0.003	0.384	0.649	1.68
	Year	0.148	0.067	0.013	0.005	0.399	0.310	0.942	0.506	0.003	0.365	0.522	1.40
After Transit of Riparian Forest													
Surface Runoff	Spring	0.056	0.136	0.084	0.029	0.131	0.124	0.560	0.053	0.008	0.176	0.188	0.425
	Summer	0.055	0.079	0.051	0.012	0.129	0.151	0.478	0.074	0.006	0.102	0.243	0.424
	Fall	0.109[b]	0.168	0.038	0.049	0.339	0.319	1.02	0.024	0.013	0.674	0.867	1.58
	Year	0.055	0.108	0.067	0.021	0.130	0.137	0.518	0.063	0.007	0.139	0.216	0.425
Groundwater	Spring	0.178	0.073	0.028	0.017	0.285	0.214	0.795	0.014	0.003	0.361	0.461	0.838
	Summer	0.158	0.072	0.023	0.035	0.270	0.161	0.719	0.016	0.007	0.375	0.408	0.806
	Fall	0.126	0.066	0.012	0.031	0.256	0.207	0.698	0.054	0.007	0.396	0.521	0.978
	Winter	0.082	0.032	0.010	0.020	0.260	0.200	0.604	0.039	0.003	0.376	0.602	1.02
	Year	0.136	0.061	0.018	0.026	0.267	0.195	0.703	0.031	0.005	0.376	0.498	0.910

[a] Most of total phosphate was orthophosphate.
[b] Only one set of samples were collected and these data were not included in the yearly average.

DISCUSSION

Ion Imbalances

One of the most interesting aspects of our data is the lack of ion balances in most instances. These discrepancies are of two types. One is the mass imbalance between ion inputs and ion outputs for a given study site. The second is the imbalance between cations and anions at a given time and place. These imbalances lead to a whole range of new hypotheses to test in future studies.

Mass Balance Imbalances

When inputs from bulk precipitation (Table I) and management of cropland (see Results, Ion Balances of Cropland section) are compared with measured outputs from the three study sites for the entire study year, a net ion gain is apparent. This is true for ions assumed to be biologically inert such as Na^+ and Cl^- as well as for total cations and anions. As mentioned earlier, this seems to be clearly due to the effects of drought. In the forest site the volume of discharge was less than 5% of precipitation. Since transpiration is higher in a forest, drought effects on ion balances are most pronounced for the forest site. Another indication that drought is the main reason for the imbalance is data taken in February, when discharge volumes were relatively high. This was the only month in which ion discharges exceeded ion inputs for all three systems. It is not clear yet how long collection of input/output ion data must continue before a reasonable mass balance is achieved. Since 1980 was also a relatively dry year, ion outputs will hopefully exceed inputs after the next period of high precipitation and thorough flushing of the soils.

Cation/Anion Imbalances

Cations and anions in bulk precipitation appear relatively well balanced for the year (Table I). However, the balance is not as good for seasonal data. Organic N in bulk precipitation included both particulate and dissolved components and was, therefore, an overestimate of dissolved or easily soluble organic amines. This is particularly troublesome in spring due to tree pollen.

All seasonal and annual watershed discharges (Table I) had significantly more measured anions than cations. The reasons include the fact that we didn't determine Fe^{2+}, Mn^{2+} or Al^{3+} (listed in order of probable importance). Although these ions were not measured routinely in this study, some data are available. Discharges of dissolved iron from four larger, mixed-land-use watersheds in 1972 were not highly different and averaged 0.0073, 0.0061, 0.0098 and 0.023 keq-ha^{-1} in the spring, summer, fall and winter, respectively, for an annual iron discharge of 0.046 keq-ha^{-1}. Year-to-year variation due to weather and variation due to differences in land use composition have not been determined. Unpublished data on manganese in stream waters indicates levels about 5–10% of those for iron. Dissolved aluminum levels in streams, including those draining the present study sites, are usually less than 0.01 meq-L^{-1}, corresponding to less than 0.002 keq-ha^{-1} of Al^{3+} discharge during the study year. The addition of these cations might have given an approximate cation/anion balance for the forest site discharges, but probably not for the managed sites.

The patterns of cation/anion balances for surface and groundwater (Tables V to VII) are more informative. In forest groundwater, for example, sulfate levels are greatly reduced at the lower elevation, suggesting biological reduction to sulfide, which was not measured in this study. In this case the sum of measured cations was more than double the sum of measured anions. This seems to make sense, but the discrepancy is about as great for surface runoff, collected at the same times. Sulfate concentrations in surface runoff were low, but always higher than those in bulk precipitation. The reduction of sulfate to sulfide in storm runoff is not very plausible, and at this time we have no suggestions of quantitatively important anions that were not measured, other than bicarbonate. Although bicarbonate was not routinely measured in these samples, in a few cases it was. In one set of surface runoff samples from pastureland, the sum of cations averaged 1.25 meq-L^{-1} and anions summed to 1.12 meq-L^{-1}. Of these anions, bicarbonate was 0.83 meq-L^{-1} or 74% of the sum. It seems that these surface runoff waters exchange their H^+ for alkaline earth cations and rapidly become neutralized since they are poorly buffered. The imbalance patterns in the groundwaters and surface waters of the pastureland site paralleled that of the forest site, except that sulfate groundwater concentrations were only slightly lower at the lower elevation. In the case of the cropland site cations anions are more nearly balanced in surface runoff (Table VII) and bicarbonate concentrations when measured were low. Groundwater at the bottom of the cropland was fairly acid (pH 4–5), probably due to the effects of nitrification. Thus, hydrogen ion would also have constituted a significant cation in the balance in addition to Fe^{2+}, Mn^{2+} and Al^{3+}.

Overview of Ion Flux Patterns in the Forest

To synthesize the dynamics of the ion flux through the forest system, simplifying schematic diagrams were drawn to illustrate important patterns of ion change (Figures 8 to 13). The spring season and a July storm were selected because they represent the most complete data available for the forest. In all of these diagrams the width of the pathway is proportional to the magnitude of the parameter measured. The percentage of total seasonal discharge occurring as quick flow (surface runoff) and slow flow (groundwater discharge) is also presented along with actual concentrations positioned relative to where the samples were taken within the system.

Average spring K^+ concentrations in surface runoff (Figure 8) were 30-fold higher than in bulk precipitation. Groundwater K^+ concentrations were 5- and 5.7-fold higher than in bulk precipitation at the upper and lower elevations, respectively. Thus, the increase in K^+ in groundwater seemed to occur primarily during infiltration. The Cl^- concentration pattern was similar qualitatively (Figure 9) to that for K^+ concentrations. Thus, much of the increase in these two ion concentra-

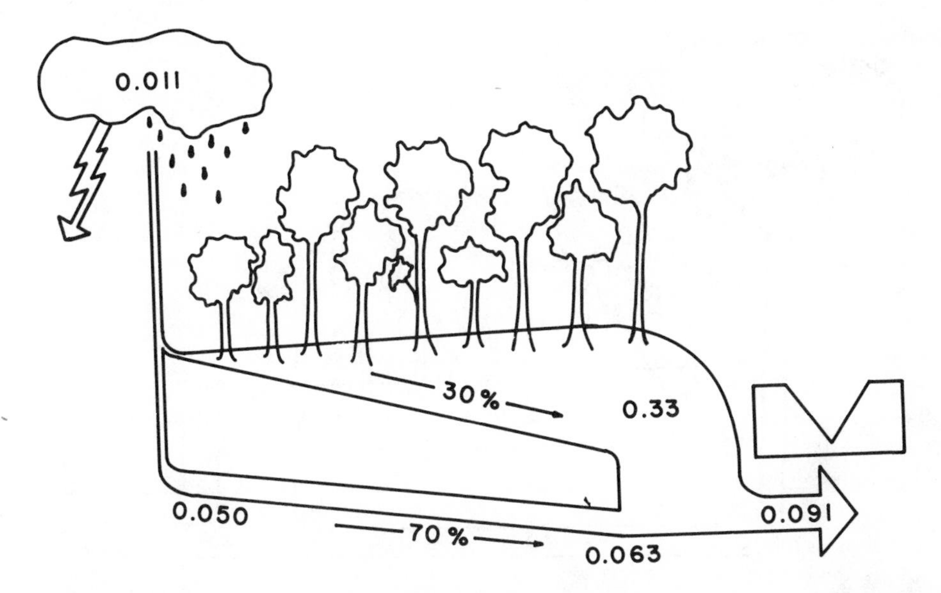

Figure 8. Diagram of average spring K^+ concentration changes (meq-L^{-1}) along different pathways through the forest study site from bulk precipitation inputs to stream outputs at the weir. Percentages given are the portions of total flow occurring as groundwater and surface runoff. Path width is proportional to the average concentration.

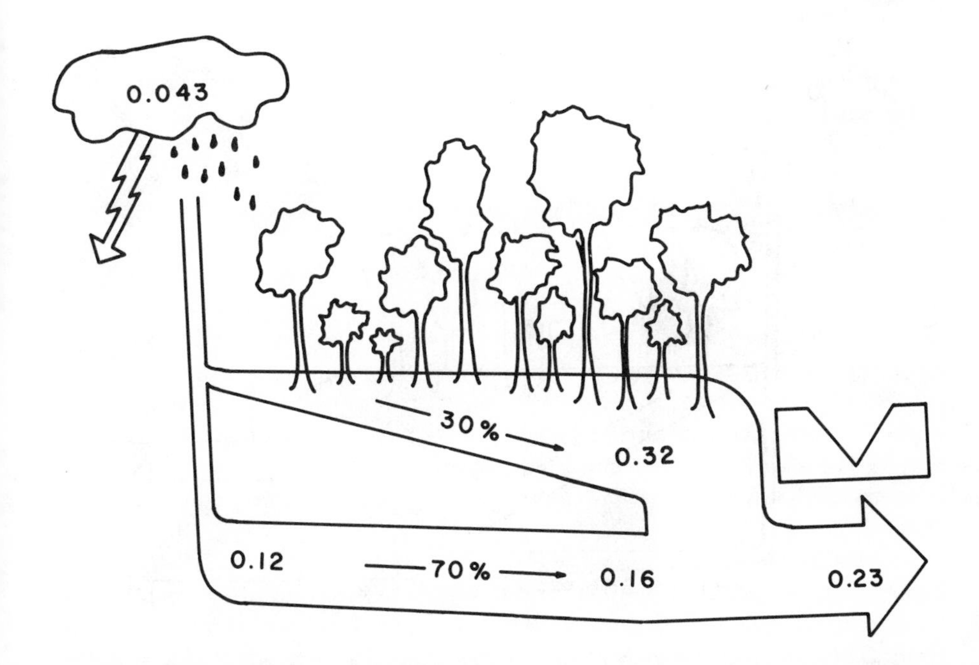

Figure 9. Diagram of average spring Cl^- concentrations (meq-L^{-1}) at the forest site. Otherwise as in Figure 8.

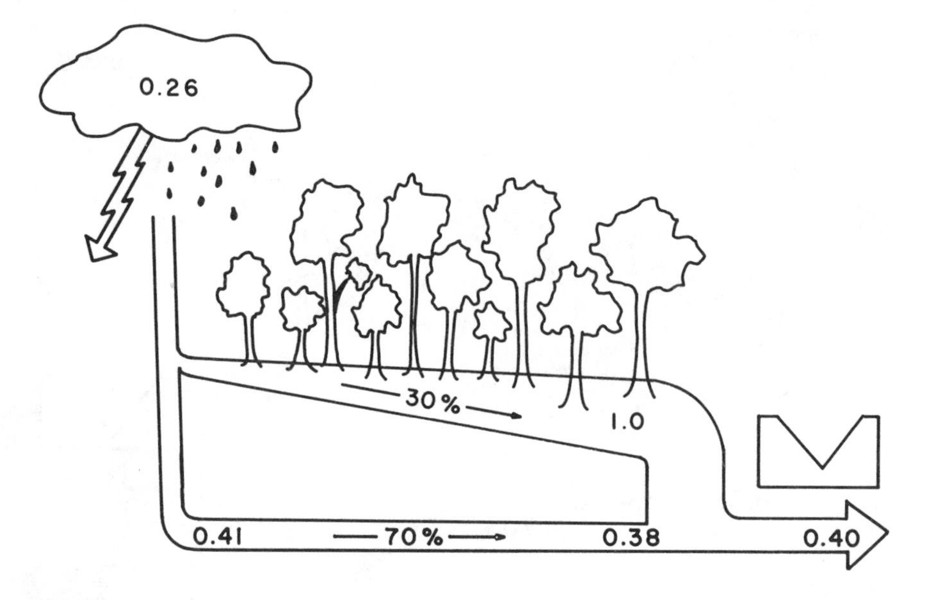

Figure 10. Diagram of average spring K^+/Cl^- ratios of the forest site. Otherwise as in Figure 8.

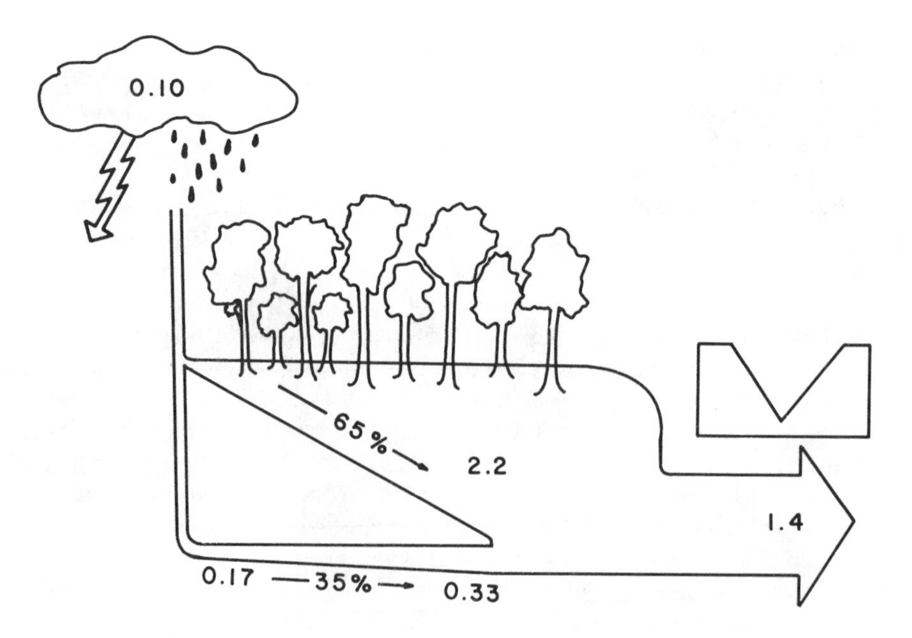

Figure 11. Diagram of K^+/Cl^- ratios during the July 4, 1981, storm week at the forest study site. Otherwise as in Figure 8.

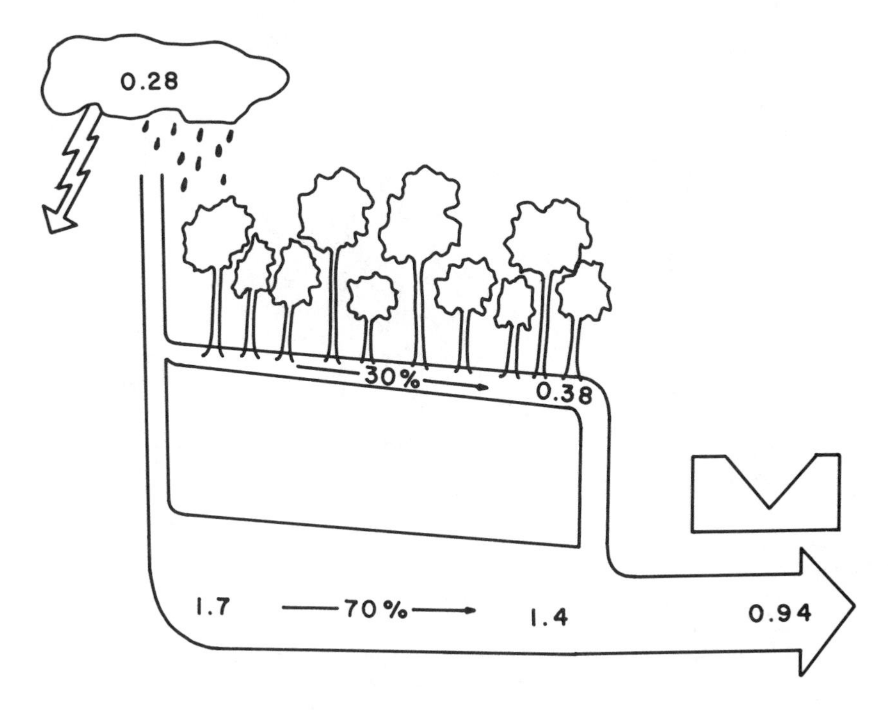

Figure 12. Diagram of average spring Mg^{2+}/Cl^- ratios at the forest site. Otherwise as in Figure 8.

tions in surface runoff was probably due to the solution from plant and litter surfaces of dry salts deposited between storms. This deposition was probably both from dryfall impaction and evapotranspiration from surface and near surface materials. Increases in the concentrations of these ions as they move down the hydrologic gradient in groundwater are probably due primarily to the concentrating effects of evapotranspiration. When the ratio of K^+ to Cl^- is diagrammed (Figure 10), the effects of evapotranspiration are minimized and a more easily interpreted pattern is evident. It is reassuring that, although all data in the diagram were obtained independently, they balance reasonably well mathematically. This balance is a check on the adequacy of the hydrograph separations into surface and groundwater discharges and also the transect approach to sampling.

An example of the K^+/Cl^- pattern for a storm event (Figure 11) is given for the July 4, 1981, storm week. In this week 65% of total volume of discharge was surface flow with a K^+/Cl^- ratio of 2.2 while 35% was groundwater flow with a ratio of 0.33 at the lower elevation. The pattern for Mg^{2+}/Cl^- ratios in the spring (Figure 12) is sharply contrasting to that for K^+/Cl^- ratios. Surface water ratios were not much higher than in bulk precipitation but groundwater ratios were much higher. An intermediate pattern was observed for the Mg^{2+}/Cl^- ratio during the May 15, 1981, storm week (Figure 13). The ratio in surface runoff was enhanced 2.6-fold over bulk precipitation that week and 1.7-fold over

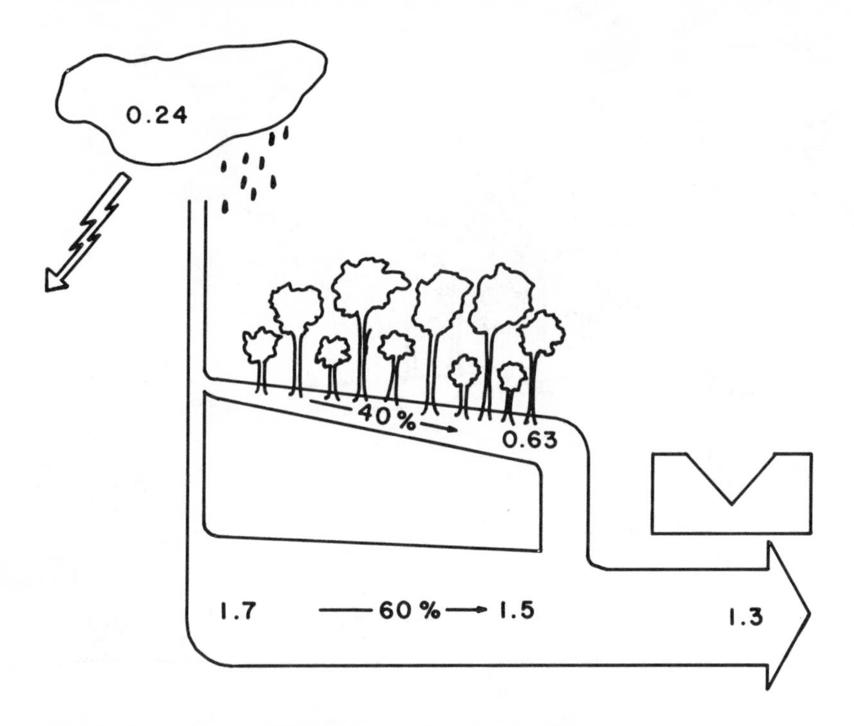

Figure 13. Diagram of Mg^{2+}/Cl^- ratios during the May 15, 1981, storm week at the forest study site. Otherwise as in Figure 8.

the average for spring surface runoff, but groundwater ratios were essentially the same. Patterns for Na^+/Cl^- and Ca^{2+}/Cl^- ratios were very similar to those for Mg^{2+}/Cl^-.

Nitrate Overview in the Cropland/Riparian System

Although nitrate fluxes into the watershed as bulk precipitation are high, they are small in comparison to inputs to cropland as fertilizer. Much of this fertilizer is applied as nitrate and ammonium. Nitrification in the surface soils oxidizes some of this to nitrate. Much of the fertilizer is also applied as organic N, often as urea. This material is also slowly mineralized and eventually yields significant nitrate discharges. Considering these large management inputs of nitrogen to croplands and the high concentrations of nitrate measured in both surface and groundwater leaving the cropland, the riparian forest seems to be carrying out an ecologically important role in the prevention of serious overenrichment of receiving waters with nitrate. The concentration pattern for nitrate during the spring in the cropland/riparian forest system (Figure 14) illustrates this beneficial role of the streamside forests. Nitrate concentrations were greatly reduced in both surface

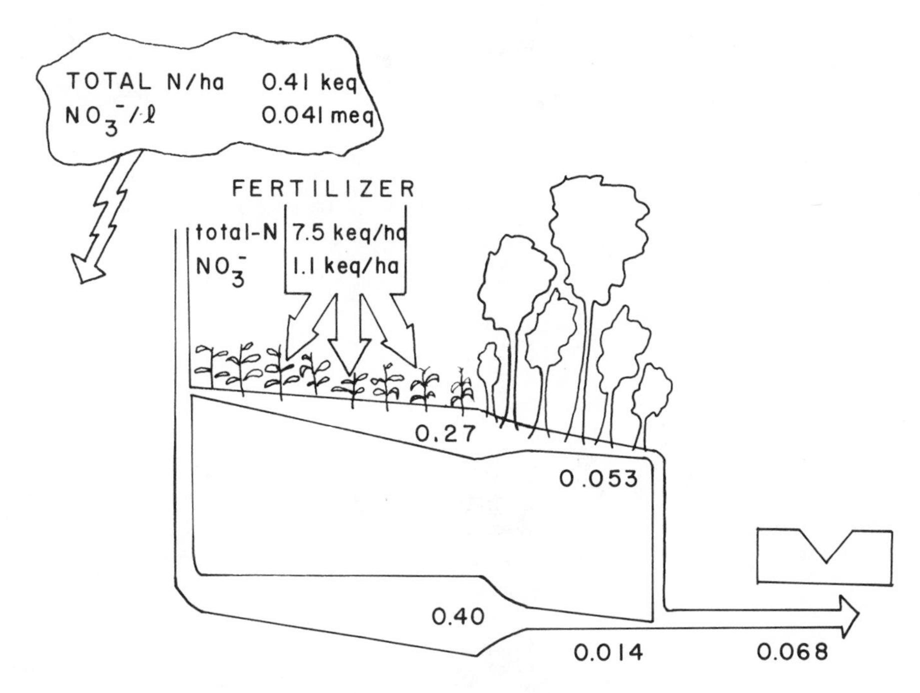

Figure 14. Diagram of average spring nitrate concentration pattern at the cropland study site, including the riparian forest. Spring fertilizer inputs were proportional relative to spring bulk precipitation inputs per hectare. Otherwise as in Figure 8.

and groundwaters during their transit of the forest. A similar pattern was observed for the other seasons. Note that the mean nitrate concentration observed at the weir was higher than the average concentration measured as surface runoff or groundwater. We believe this is due to areas within the riparian forest where conditions are not favorable for significant denitrification.

Comparisons with Other Studies

Bulk precipitation and forest discharge data from this study are compared with data from two other study sites in Table VIII. Hydrogen ion inputs were about equal at the Hubbard Brook site in New Hampshire but Na^+ and Cl^- input fluxes were about half of those at Rhode River. Acidity of precipitation at the Como Creek site in Colorado was much less, Na^+ flux intermediate when compared to the other two sites, and Cl^- was not measured. Ion ratios in bulk precipitation for K^+ and Mg^{2+} were about the same at Rhode River and Hubbard Brook, while Ca^{2+}/Cl^- ratios were higher at Hubbard Brook. At Como Creek, the K^+/Na^+ ratio was somewhat higher, Mg^{2+}/Na^+ lower, and Ca^{2+}/Na^+ was much higher. When ion ratios in bulk precipitation are compared with the same ratios in land discharges (Table VIII), it is apparent that while the K^+/Na^+ ratio declined at

Table VIII. Comparisons of Ion Fluxes Between Selected Mature Forest Sites on Noncalcareous Soils (Units are keq-ha^{-1}-y^{-1} or Ratio of Equivalents)

Site	H^+	Na^+	Cl^-	K^+/Na^+	K^+/Cl^-	Mg^{2+}/Na^+	Mg^{2+}/Cl^-	Ca^{2+}/Na^+	Ca^{2+}/Cl^-
Annual Bulk Precipitation Inputs									
Rhode River	0.94	0.12	0.32	0.35	0.13	0.70	0.27	0.88	0.33
Hubbard Brook [5,6]	0.92	0.067	0.17	0.34	0.13	0.63	0.25	1.5	0.60
Como Creek [10]	0.11	0.086	[a]	0.48	[a]	0.43	[a]	2.3	[a]
Land Discharge Outputs									
Rhode River	0.0021	0.072	0.080	0.45	0.41	1.3	1.2	0.87	0.79
Hubbard Brook [5,6]	0.012	0.28	0.12	0.14	0.31	0.82	1.9	1.7	3.9
Como Creek [10]	0.00018	0.069	[a]	0.12	[a]	0.78	[a]	1.7	[a]

[a] Chloride was not measured in the Como Creek study.

the other two sites, it increased at the Rhode River site. The K^+/Cl^- ratio is more useful for comparison since sodium itself probably is being displaced from these systems. This ratio increased 2.4-fold at Hubbard Brook and 3.2-fold at Rhode River. The Mg^{2+}/Na^+ ratio increased in land discharges at all three sites and the Mg^{2+}/Cl^- ratio increased more than the K^+/Cl^- ratio at both Rhode River and Hubbard Brook. The Ca^{2+}/Na^+ ratio was relatively constant, except at Como Creek, where it declined. The Ca^{2+}/Cl^- ratio more than doubled in forest land discharges at Rhode River and increased over sixfold at Hubbard Brook. Thus, based on ion ratios to chloride, both the Hubbard Brook and Rhode River systems appear to be displacing K^+, Ca^{2+} and Mg^{2+}. Judging from sodium ratios, Como Creek is displacing less K^+ and Ca^{2+}. If true, this might be explained by the differences in H^+ inputs (Table VIII).

Anticipated Biological and Ecological Effects

If K^+, Mg^{2+} and Ca^{2+} are being displaced from the plant/soil system by H^+ in precipitation, one would like to have a temporal perspective of how rapidly this process will deplete the system of these essential plant nutrients. Especially for the forest system, how much is the annual net outward flux as compared to the available nutrient pool? To calculate a first approximation we assumed that the annual forest land discharge of these ions would, on the average, equal the product of the precipitation loading measured in this study and the proportional increase in discharges over bulk precipitation in the ion's ratio to Cl^-. For example, K^+/Cl^- ratios in forest discharges were 3.1-fold higher than in bulk precipitation (Table I). Therefore, average annual K^+ output was estimated to be 3.1 times 0.042 (the annual K^+ loading from precipitation) or 0.13 keq-ha^{-1}-y^{-1}. As a check on this calculation one would expect that H^+ fluxes into the system should approximately equal the sum of estimated annual outputs of Na^+, K^+, Mg^{2+} and Ca^{2+}. These are 0.288, 0.130, 0.382 and 0.264 keq-ha^{-1}-y^{-1} for Na^+, K^+, Mg^{2+} and Ca^{2+}, respectively. The net estimated output of these ions by this method is 1.06 keq-ha^{-1}-y^{-1}. This is not very different from the bulk precipitation inputs of 0.94 keq-ha^{-1}-y^{-1} of hydrogen ion. Thus, it seems reasonable to use these as an approximate rate of loss on an "average" year.

In a forest system most of the nutrient mass is usually stored in woody plant biomass and forests on the Rhode River site are no exception. The real issue, however, is how much of these plant nutrient elements are found in the soil. Relevant data on the composition of the soils of the forest study site is summarized in Table IX. Exchangeable potassium and magnesium were extracted with 1 *N* ammonium acetate. Total calcium and exchangeable potassium and magnesium were then determined by digestion with nitric acid, addition of lanthanum nitrate and atomic absorption spectroscopy. Appropriate controls, standards and spiked samples were also analyzed in order to correct for matrix effects. From these data (Table IX) it is apparent that there is a fairly large pool of exchangeable magnesium in the surface soils. However, the pools of potassium and calcium in

Table IX. Pools of Potassium, Magnesium and Calcium in Surface Litter and Surface Soils of Mature Forest Site at Rhode River in 1976

Zone	*Dry wt. (kg-m⁻²)*	*Exchangeable K^+*		*Exchangeable Mg^{2+}*		*Total Ca^{2+}*	
		$mg\text{-}g^{-1}$	*$keq\text{-}ha^{-1}$*	*$mg\text{-}g^{-1}$*	*$keq\text{-}ha^{-1}$*	*$mg\text{-}g^{-1}$*	*$keq\text{-}ha^{-1}$*
Leaf litter	1.14	0.25	0.073	2.8	2.66	1.5	0.86
Soil							
0–3 cm	24.8	0.25	1.59	2.1	43.4	0.08	0.99
3–5 cm	27.4	0.24	1.69	1.8	41.1	0.07	0.96
5–8 cm	39.8	0.16	1.63	1.5	49.8	0.06	1.19
8–12 cm	61.1	0.17	2.79	1.4	74.8	0.05	1.60
12–18 cm	93.9	0.17	4.09	1.7	133	0.10	4.70
18–24 cm	94.4	0.17	4.11	2.1	165	0.14	6.61
Sum	342		16.0		510		16.9

surface litter and the first few centimeters of soil, in which root hairs are most active for nutrient uptake, are rather small. They are much larger than the estimated annual flux rate, but could conceivably be completely depleted in a few years or decades, depending on the depth zone one considers. Indeed, this is a conservative estimate, because the rate of H^+ input and hence of cation output has been increasing over the past seven years (Figure 2). Also, much of the nutrient pool currently observed in the soil may really be either part of the root and microbial biomass in the soil or have been recently displaced from above ground and litter pools.

Plants require more calcium than any other mineral element and moderate levels of potassium for growth. Depletion of the available supply of these elements from forest systems is a matter of considerable concern, especially when surface soils do not contain these nutrients in the parent soil minerals. Native forest species have been selected over time for their ability to compete effectively for these ions and to keep them tightly recycled. Also, some species require more of a given element to survive. With hydrogen ion inputs from precipitation increasing rapidly, many plant species may no longer be able to successfully compete for these elements. Thus, the gradual loss of key elements such as calcium may, over time, bring about significant shifts in the species composition of forests [25]. In the case of calcium and potassium, the shorter-term effects anticipated include decreased sexual reproduction, a decline in general vegetative physiological vigor, and a consequent increase in the incidence of disease and insect attack [26,27].

CONCLUSIONS

The results from this study confirm others which have found a trend of increasing acidity in rainfall and important regional differences in its effects. It is also relatively clear that at the Rhode River site, increased hydrogen ion inputs are displacing the essential plant nutrients of Mg^{2+}, Ca^{2+} and K^+. Although displacement rates are apparently low, available pools in forested areas could be depleted in a few decades and cause ecologically significant effects.

Ion losses appeared to be proportional to the magnitude of disturbance associated with the three land uses studied. Thus total cation and anion outputs were lowest from the forest site, similar but higher for the pastureland, and significantly greater in the cropland discharge than in either of the other two. Concentrations at the weirs closely approximated the discharge-weighted concentrations of surface and groundwater, indicating that analysis of these different flow pathways through a watershed is an important key to understanding the origins of the final output concentrations.

The functional importance of the riparian forest in reducing nitrate concentrations in discharge from an agricultural watershed was clearly shown and raises interesting questions as to the generality of this result. Other questions raised by this study are the importance of Fe, Mn, and Al ions in intrawatershed patterns of ion change and what measures should be taken to best compensate for K^+, Mg^{2+} and Ca^{2+} losses. Most importantly, however, this study shows how the concentrations of various chemical components in stream waters are partially controlled by the complex interplay of chemical factors (such as pH), physical/biological factors (such as the redox potential), climatic factors (such as droughts and intense storms) and anthropogenic factors (such as land use and acid rain). All of these factors when taken separately are not perfectly understood, let alone when they are allowed to interact. Thus, future research along these lines is both needed and essential to achieve a better understanding of multiple land use systems.

ACKNOWLEDGMENTS

This research was supported in part by NSF grant DEB-79-11563 and by the Smithsonian Environmental Sciences Program. It is also part of a master's thesis submitted to Miami University.

REFERENCES

1. Cogbill, C.V. "The History and Character of Acid Precipitation in Eastern North America," *Water, Air, Soil Poll.* 6(2/3/4):407–413 (1976).
2. Cowling, E.B., and R.A. Linthurst. "The Acid Precipitation Phenomenon and Its Ecological Consequences," *Bioscience* 31(9):649–653 (1981).
3. Malmer, N. "Acid Precipitation: Chemical Changes in the Soil," *Ambio* 5(5–6):231–234 (1976).
4. Abrahamsen, G., K. Bjor and B. Tveite. "Effects of Acid Precipitation on Coniferous Forest," in *Impact of Acid Precipitation on Forest and Freshwater Ecosystems in Norway,* F. H. Braekke, Ed. (Oslo, Norway: SNSF Project, 1976), pp. 31–63.

5. Bormann, F.H., G.E. Likens and J.S. Eaton. "Biotic Regulation of Particulate and Solution Losses from a Forest Ecosystem," *Bioscience* 19(7):600–610 (1969).
6. Likens, G.E., F.H. Bormann, J.S. Eaton, R.S. Pierce and N.M. Johnson. "Hydrogen Ion Input to the Hubbard Brook Experimental Forest, New Hampshire, During the Last Decade," *Water, Air, Soil Poll.* 6(2/3/4):435–445 (1976).
7. Martin, C.W. "Precipitation and Streamwater Chemistry in an Undisturbed Forested Watershed in New Hampshire," *Ecology* 60(1):36–42 (1979).
8. Johnson, P.L., and W.T. Swank. "Studies of Cation Budgets in the Southern Appalachians on Four Experimental Watersheds with Contrasting Vegetation," *Ecology* 54(1):70–80 (1973).
9. Swank, W.T., and G.S. Henderson. "Atmospheric Input of Some Cations and Anions to Forest Ecosystems in North Carolina and Tennessee," *Water Resources Res.* 12(3):541–546 (1976).
10. Lewis, W.M. Jr., and M.C. Grant. "Changes in the Output of Ions from a Watershed as a Result of the Acidification of Precipitation," *Ecology* 60(6):1093–1097 (1979).
11. Gosz, J.R. "Nutrient Budget Studies for Forests Along an Elevational Gradient in New Mexico," *Ecology* 61(3):515–521 (1980).
12. Ehrlich, P.R., A.H. Ehrlich and J.P. Holdren. *Ecoscience: Population, Resources, Environment* (San Francisco, CA: W.H. Freeman, 1977), p. 252.
13. Correll, D.L., and D. Ford. "Comparison of Precipitation and Land Runoff as Sources of Estuarine Nitrogen," *Est. Coast. Shelf Sci.* (in press).
14. Miklas, J., T.L. Wu, A. Hiatt and D.L. Correll. "Nutrient Loading of the Rhode River Watershed via Land Use Practice and Precipitation," in *Watershed Research in Eastern North America, Vol. I,* D.L. Correll, Ed. (Edgewater, MD: Chesapeake Bay Center for Environmental Studies, 1977), pp. 169–194.
15. Correll, D.L. "Nutrient Mass Balances for the Watershed, Headwaters Intertidal Zone, and Basin of the Rhode River Estuary," *Limnol. Oceanog.* 26(6):1142–1149 (1981).
16. Chirlin, G.R., and R.W. Schaffner. "Observations on the Water Balances for Seven Sub-basins of Rhode River, Maryland," in *Watershed Research in Eastern North America, Vol. I,* D.L. Correll, Ed. (Edgewater, MD: Chesapeake Bay Center for Environmental Studies, 1977), pp. 277–306.
17. Higman, D. *An Environmental History of Southern Anne Arundel County, Maryland, from Prehistory Through the Nineteenth Century, with Emphasis on the Rhode River Watershed Area* (Lanham, MD: University Press of America, in press).
18. Roberts, L.M. "A Study of Erosion and Sedimentation on a Small Maryland Watershed," MS Thesis, University of Maryland (1979).
19. Correll, D.L., and D. Dixon. "Relationship of Nitrogen Discharge to Land Use on Rhode River Watersheds," *Agro-Ecosyst.* 6(2):147–159 (1980).
20. Martin, D.F. *Marine Chemistry, Vol. I* (New York: Marcel Dekker, 1972), p. 173.
21. Richards, F.A., and R.A. Kletsch. "The Spectrophotometric Determination of Ammonia and Labile Amino Compounds in Fresh and Seawater by Oxidation to Nitrite," in *Sugawara Festival Volume* (Tokyo, Japan: Maruza Co., Ltd, 1964), pp. 65–81.
22. *Standard Methods for the Examination of Water and Waste Water, 14th ed.* (New York: American Public Health Association, 1975), p. 479.
23. Higman, D., and D.L. Correll. "Seasonal and Yearly Variation in Meteorological Parameters at the Chesapeake Bay Center for Environmental Studies," in *Environmental Data Summary for the Rhode River Ecosystem (1970–1978), Section A: Long-Term*

Physical/Chemical Data, Part I: Airshed and Watershed, D.L. Correll, Ed. (Edgewater, MD: Chesapeake Bay Center for Environmental Studies, in press), pp. 1–159.

24. Barnes, B.S. "Discussion on Analysis of Runoff Characteristics by O.H. Meyers," *Trans. Am. Soc. Civil Eng.* 105:104–106 (1940).
25. Jefferies, R.L., D. Laycock, G.R. Stewart and A.P. Sims. "The Properties of Mechanisms Involved in the Uptake and Utilization of Calcium and Potassium by Plants in Relation to an Understanding of Plant Distribution," in *Ecological Aspects of the Mineral Nutrition of Plants, Vol. I,* H. Rorison, Ed. (Oxford, England: Blackwell Scientific, 1968), pp. 281–307.
26. Marschner, H. "Calcium Nutrition of Higher Plants," *Neth. J. Agric. Sci.* 22: 275–282 (1974).
27. Gauch, H.G. *Inorganic Plant Nutrition* (Stroudsburg, PA: Dowden, Hutchinson and Ross, 1972), p. 334.

CHAPTER 6

Regional Assessment of Sensitivity to Acidic Deposition for Eastern Canada

A.E. Lucas
D.W. Cowell

Acidic deposition is a long-range phenomenon affecting extensive areas of North America. The possible deleterious effects of anthropogenic acidic deposition depend on the nature of the ecosystem and chemistry of the precipitation. The delineation of "sensitive" and "susceptible" areas, on a regional scale, is considered necessary for a number of purposes. By presenting an overview, outlining the extent and distribution of problem areas, the selection of study sites for more detailed ecological research may be simplified. Regional sensitivity assessments combined with measured input loadings can be used to project the impact of acidic deposition on the regional economy [1]. This, in turn, can be used to provide part of the basis for emission control strategies.

In recent years a number of regional sensitivity assessments have been described in the literature [2–6]. To date, most of these have focused on specific target ecosystem components (such as soils, bedrock geology or forest species) without regard for ecosystem interactions or combined effects [7]. The need for a holistic approach to sensitivity assessment has been expressed by a number of authors [8–10]. Although it is not possible to model all ecosystem processes, some consideration should be given to those factors that, in combination, best express an ecosystem sensitivity. Single-factor sensitivity analyses may be suitable for some purposes; however, they can be misleading if interpreted incorrectly. For example, in many areas soils do not reflect the underlying bedrock lithology. In glaciated regions, such as northern Ontario, carbonate-rich surficial material may overlie granitic bedrock. Interpretations based on either the soil or bedrock alone would be inaccurate [5].

This chapter presents a regional sensitivity assessment of surface waters in eastern Canada. Lakes within areas identified as having a high sensitivity are the most likely to become acidified. The interpreted sensitivity is based on the potential for soils and bedrock, in combination, to reduce the acidity of acidic deposition before its entry into the aquatic regime. The approach taken is a refinement of the technique outlined by Cowell et al. [7].

SENSITIVITY OF AQUATIC SYSTEMS TO ACIDIC DEPOSITION

Acidic deposition reaches surface waters either directly from the atmosphere or indirectly through the terrestrial ecosystem. Indirect inputs include surface runoff and groundwater, both of which are in contact with vegetation, soil and, to varying degrees, bedrock. Due to the relative surface areas of land and water within the watershed, surface water chemistry is largely influenced by terrestrial factors. Sensitivity evaluations based on parameters such as lake and stream pH, alkalinity or calcite saturation index [8,11–15] are a useful indication of acidification at any one time. They do not, however, necessarily reflect the long-term capacity of watersheds to neutralize acidic deposition [1].

Ideally, aquatic chemistry in combination with terrestrial geochemistry and input loadings should be evaluated to derive a better understanding of sensitivity and susceptibility. Regional assessments using county or other nonwatershed frameworks tend to overgeneralize the relationships between water and land components. Initial attempts at relating aquatic and terrestrial factors on a regional scale have been carried out with limited success [16]. These relationships are best defined on a watershed basis.

CRITERIA FOR SENSITIVITY ASSESSMENT

In this analysis "sensitive" aquatic systems are those that occur in areas having little or no capacity to reduce the acidity of atmospheric deposition. Three levels of sensitivity have been assigned to individual map units by combining specific criteria (Table I). Limits for each parameter were assigned a high, moderate or low potential to reduce acidity based primarily on previous single-factor assessments. Terrestrial factors include soil chemistry (exchangeable bases or its surrogates of particle size, texture, pH and cation exchange capacity (CEC), and sulfate adsorption capacity), soil depth and drainage, present vegetation cover and type, topographic relief and bedrock geology. The limits of the three classes for exchangeable base content are those of Wang and Coote [6]. The basis for the description of the bedrock classes was provided by sensitivity estimates for different bedrock lithologies [2,5]. Criteria limits for other factors are more qualitative, based largely on inferred relationships between each criterion and its potential to modify surface water chemistry.

DATA BASE COMPILATION AND MAPPING CRITERIA

Three parameters have been incorporated into the sensitivity interpretation for eastern Canada. These are bedrock lithology, soil depth (including percent exposed bedrock) and soil chemistry (or available surrogates). The soil and bedrock factors are largely the last terrestrial ecosystem components to interact with acidic deposition before its entry into the aquatic regime. It is assumed that water chemistry

Table I. Terrestrial Factors and Associated Criteria for Determining the Potential of Terrestrial Ecosystems to Reduce the Acidity of Atmospheric Deposition (Modified After Cowell et al. [7])

Terrestrial Factors	*Potential to Reduce Acidity of Atmospheric Deposition*		
	High	*Moderate*	*Low*
Soil Chemistry			
Exchangeable Bases (meq-100 g^{-1}) Surrogates	>15	6–15	<6
Family Particle Size and pH in water	Clayey, pH >5.0; loamy, pH >5.5; all calcareous soils	Clayey, pH 4.5–5.0; loamy, pH 5.0–5.5; sandy, pH >5.5	Clayey, pH <4.5; loamy, pH <5.0; sandy, pH <5.5
Texture	Clay, silty clay, sandy clay (>35% clay)	Silty clay loam, clay loam, sandy clay loam, silt loam, loam (10 to 35% clay)	Silt, sandy loam, loamy sand, sand (<10% clay)
Cation Exchange Capacity (meq-100 g^{-1})	>25	10 to 25	<10
SO_4^{2-} Adsorption Capacity	High sulfate adsorption and low organic matter, and high Al_2O_3 and/or $Fe_2O_3 + Fe_3O_4$ content		Low sulfate adsorption and high organic matter, and/or low Al_2O_3 and/or $Fe_2O_3 + Fe_3O_4$ content
Soil Depth (cm)	>100	25–100 cm	<25
Soil Drainage	Poor	Imperfect to well	Rapid
Landform Relief	Level	Rolling	Steep
Vegetation	Deciduous	Mixed	Coniferous
Vegetation Cover	Continuous (>60%)	Discontinuous to sparse (20–60%)	Sparse to barren (<20%)
Underlying Material			
Parent Material	Carbonate bearing	Noncarbonate bearing	Noncarbonate bearing
Bedrock Material	Limestone, dolomite, and metamorphic equivalents, calcareous clastic rocks, carbonate rocks interbedded with noncarbonate rocks	Volcanic rocks, shales, greywackes, sandstones, ultramafic rocks, gabbro, mudstone, metaequivalents	Granite, granite gneiss, orthoquartzite, syenite

will reflect the combined effects of these factors. Simple basin flow conditions are also assumed and no consideration has been given to soil or rock groundwater residence times or deep groundwater contributions. Vegetation, relief and drainage have not yet been incorporated into the analysis.

The sensitivity map of eastern Canada (Figure 1) was originally compiled at a scale of 1:1,000,000. It is reproduced here at 1:15,000,000 with considerable loss of detail. The information used to produce the sensitivity map varies in terms of quality and quantity of data, within and between provinces. In all cases, however, they represent the best data readily available for regional interpretation. The six data sources used in the analysis are listed in Table II [5,17–23]. The sensitivity evaluations were based primarily on information interpreted within two mapping frameworks; the Ecodistrict Data Base and the Ontario Land Inventory. Ecodistricts were the major mapping unit in northern Ontario, Quebec and the Atlantic provinces where published data, at suitable scale, were not available. According to the Canada Committee on Ecological Land Classification, ecodistricts are defined as units of land "characterized by a distinctive pattern of relief, geology, geomorphology, vegetation, soils, water and fauna" as mapped at scales between 1:125,000 and 1:500,000 [24].

In Ontario, south of 52° N latitude the interpreted map units are based on the Ontario Land Inventory (OLI) Classification polygons. The Ontario Land Inventory Classification mapping base (published at 1:250,000) contains data relating to soil texture, depth, drainage, and depth to carbonate [25].

Polygons within the OLI framework have an average size of 45 km^2, whereas ecodistricts may be as large as 35,000 km^2. The differences in spatial resolution are reflected in both the data criteria and classes used in the interpretation and in the final sensitivity map (Figure 1).

Bedrock interpretations for eastern Canada south of 52° N latitude were based on Shilts et al. [5]. In northern Ontario the Ontario Geological Survey maps (Ontario Ministry of Natural Resources, Maps 2200 and 2201) supplied the required bedrock information. Both of these data sets were published at 1:1,000,000. Elsewhere, bedrock data already compiled within the Ecodistrict Data Base were used.

Although information pertaining to soil depth was similar throughout eastern Canada, soil chemical data were sparse outside of agricultural areas. An estimation of the potential of soils to reduce acidity is derived from textural information contained within the Ecodistrict Data Base. A low potential is inferred for sandy soils and a high potential is equated with clay-sized material. Although soil texture is not an ideal surrogate for exchangeable base content, textural data are the only information available for regional analyses for Quebec and the Atlantic provinces.

Within the framework of the OLI, depth to carbonate is, perhaps, a better estimate of soil chemistry. While it is based on field testing (with HCl), field sites were limited, although widely distributed. Between test sites, information was interpreted through known relationships with textural classes. The high lime class indicates lime near the surface of the soil profile. This has been equated to a high potential to reduce acidity. A low potential to reduce acidity is interpreted from the "no lime" class.

In northern Ontario textural data from the Ecodistrict Data Base were supple-

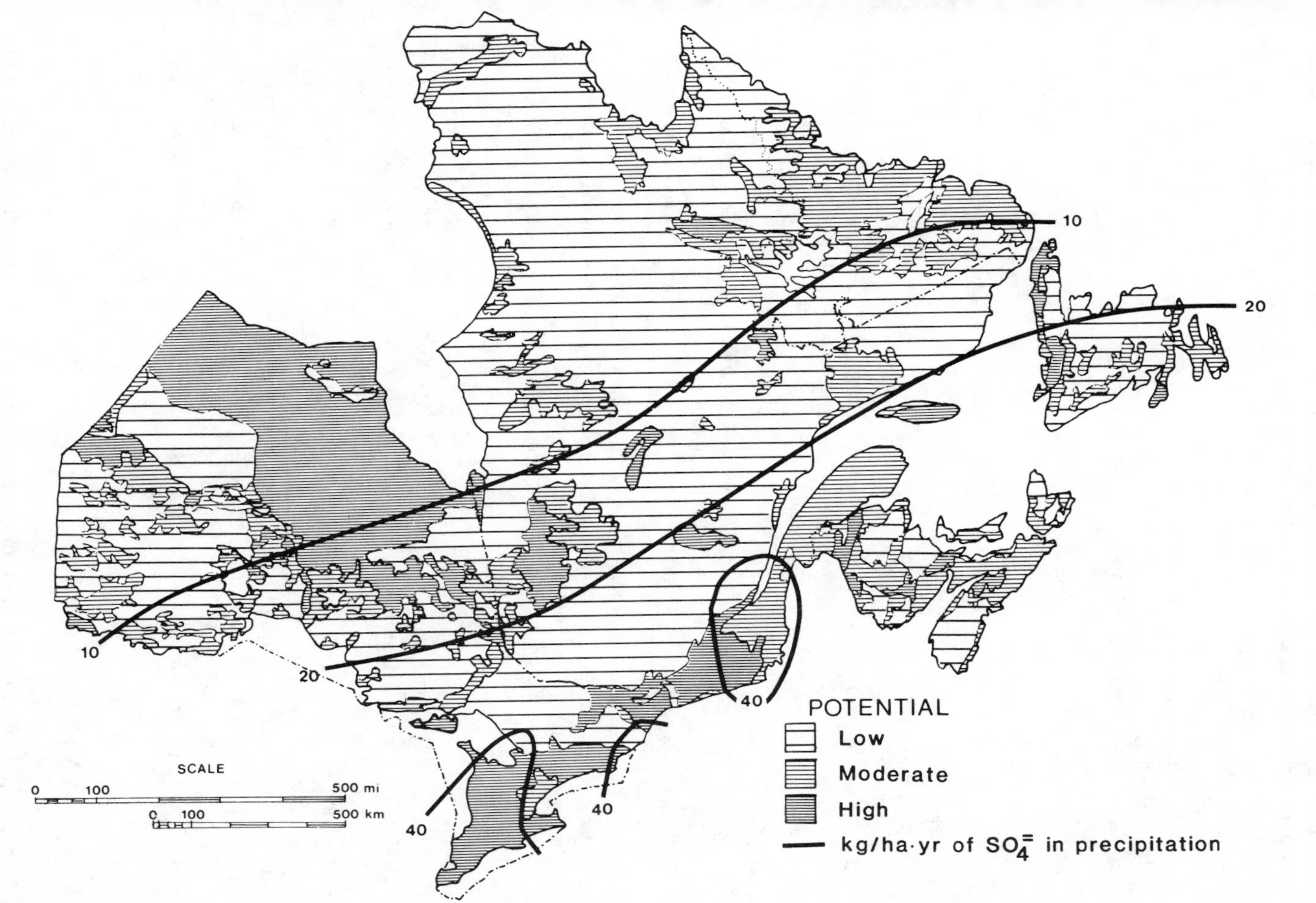

Figure 1. Potential of soils and bedrock to reduce the acidity of atmospheric deposition in eastern Canada. Contours show wet sulfate deposition as measured at CANSAP stations.

Table II. Terrestrial Factors Used for the Interpretation of the Potential to Reduce Acidity of Atmospheric Deposition in Eastern Canada

Terrestrial Criteria and Classes	*Coverage*	*Reference*
Bedrock Geology		
High, moderate, low sensitivity	Ontario, Quebec, Maritimes, Newfoundland	5
Lithology	Northern Ontario	Ontario Geological Mapping 17 (MNR Maps 2200, 2201)
	Northern Quebec, Labrador	Ecodistrict Data Base 18, 19
Soil Depth		
Deep, shallow, very shallow, bare	Ontario	Ontario Land Inventory 20
Deep, shallow, bare	Northern Ontario, Quebec, Maritimes, Newfoundland/Labrador	Ecodistrict Data Base 18, 19, 21
Soil Chemistry Surrogates		
Texture (clay, loam, sand)	Quebec, Maritimes, Newfoundland/Labrador	Ecodistrict Data Base 18, 19
Depth to carbonate (high, low, no lime)	Ontario	Ontario Land Inventory 20
Glacial Landforms	Northwestern Ontario	22
Organic Soils (≥50% mapping unit)	Northern Ontario	Wetlands Working Group (1981) 23
	Quebec, Maritimes, Newfoundland/Labrador	Ecodistrict Data Base 18, 19

mented with information from physiographic maps [22] and the wetlands map of Canada [23]. These were compiled at 1:500,000 (1 inch to 90 miles) and 1:7,500,000 (1 inch to 120 miles), respectively, and provided a much needed refinement.

In comparison with the smaller-scale soil map of Canada [26] there appears to be good correlation between texture or depth to carbonate and soil order. Sand or no lime soils correspond to acid podzols (or "Rockland") and clay or high lime soils are predominantly luvisols (clay-rich) and gleysols (seasonally wet, poorly drained). Areas with loam or low lime soils are more varied and include brunisols, regosols (poorly developed), luvisols and podzols [27]. Areas identified as predominantly peatland are mapped within the OLI and Ecodistrict Base as organic soils [26].

MAPPING METHODOLOGY

To assign map polygons a high, moderate or low potential to reduce acidity, each of the three variables needs to be evaluated individually. The number of potential combinations is virtually unlimited when the percent of deep and shallow soil and exposed bedrock is considered for each map unit. Thus, certain assumptions need to be made to standardize the decision-making process and minimize the number of possible combinations.

The flow diagram presented in Figure 2 illustrates the decision-making process. It is a conceptual framework based on inferred relationships between combinations of factors and the overall ability of the map unit to reduce acidity. Altering the basic assumptions and boundary values (e.g., percent bedrock exposure) would affect the final map product. Boundary values (50 and 75%) were assigned arbitrarily to produce a manageable number of classes. Their significance has not yet been validated. It is assumed that all drainage waters in carbonate-rich terrains contact bedrock (case 1 in Figure 2). Therefore, all areas underlain by limestone, dolomite or other carbonate-rich bedrock (Table I) are assigned a high potential to reduce acidity, regardless of the overlying soil characteristics.

The second assumption (case 11, Figure 2) identifies areas dominated by organic soils (≥50% of map polygon). Because these areas already contribute low-pH, low-bicarbonate water to the watershed, they have a low potential to reduce the acidity of atmospheric deposition. The only exception to this occurs in the case of organic terrain underlain by carbonate bedrock (such as in the Hudson Bay Lowland). At this time it is not known to what degree peatlands and organic groundwater are affected by, or in turn modify, incoming anthropogenic mineral acids.

Soil depth is used in the next stage of the interpretation. Where bedrock exposure equals or exceeds 50% of the map polygon, the emphasis is on the bedrock capacity to reduce acidity (cases 2, 3 and 4). Case 5, Figure 2, is an exception to this. The combination of deep or shallow soils of high potential overlying bedrock of low potential with 50–74% exposed are averaged and assigned a moderate potential. In this instance, it is assumed that although the soils can substantially reduce the acidity they do not cover a sufficient area to warrant an overall assessment of high potential. This is countered by the low potential of the predominantly exposed bedrock.

Interpretation of map units dominated by shallow soils is based on the combined potentials of bedrock and soil characteristics (cases 6, 7 and 8). Where shallow soils of high potential overlie bedrock of moderate potential, the polygon is assigned an overall assessment of high potential to reduce acidity (case 9). Soil properties define the potential for all map units dominated by deep soils (case 10).

RESULTS

Specific combinations of factors that occur in eastern Canada are shown in Table III, according to their assessed potential to reduce acidity. The distribution of high, moderate and low classes is illustrated in Figure 1. The map was compiled

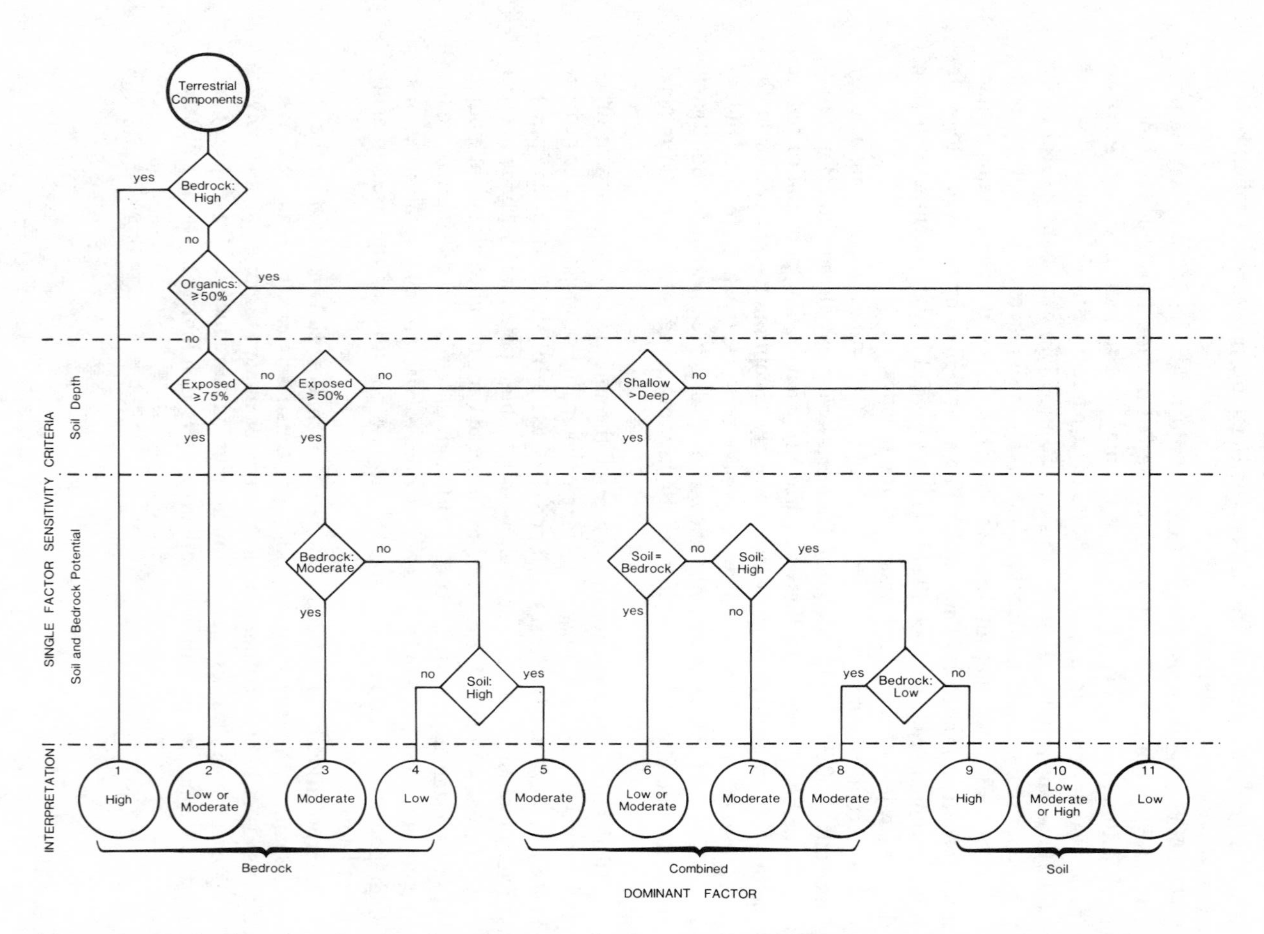

Figure 2. Evaluation procedure for the assessment of soil and bedrock potential to reduce acidity of atmospheric deposition.

Table III. Terrestrial Characteristics of Areas Having High, Moderate and Low Potential to Reduce Acidity for Eastern Canada

Potential to Reduce Acidity		*Soil and Bedrock Characteristics*[a]	*Case (From Figure 2)*
High	H1	All areas underlain by bedrock having a high potential (e.g., limestone) regardless of outcropping	1
	H2	Shallow clay or high lime soils overlying bedrock having a moderate potential to reduce acidity	9
	H3	Deep clay or high lime soils	10
Moderate	M1	Bedrock having moderate potential exposed in at least 50% of area	2,3
	M2	Clay or high lime soils overlying bedrock having a low potential with 50 to 74% outcropping	5
	M3	Shallow loam or low lime soils overlying bedrock of low potential	7
	M4	Shallow sand or no lime soils overlying bedrock of moderate potential	7
	M5	Shallow clay or high lime soils overlying bedrock of low potential	8
	M6	Shallow loam or low lime soils overlying bedrock having moderate potential	6
	M7	Deep loam or low lime soils	10
Low	L1	Bedrock having low potential exposed in at least 75% of area	2
	L2	Deep or shallow sand or loam (no lime or low lime) soils overlying bedrock having a low potential with 50 to 74% outcropping	4
	L3	Shallow sand or no lime soils overlying bedrock having low potential	6
	L4	Deep sand or no lime soils	10
	L5	At least 50% organic soils underlain by bedrock having low or moderate potential	11

[a] Unless otherwise specified, percent exposed bedrock is assumed to be less than 50%.

at 1:1,000,000. Figure 1, at 1:15,000,000, is a generalization of the original interpretation with much of the detail missing.

High Potential to Reduce Acidity

Map units with a high potential to neutralize acidic deposition are predominantly underlain by carbonate bedrock (H1) and areas dominated by deep clay or high lime soils (H3). These cover approximately 16% and 4% (Table IV) of the map area represented in Figure 1, respectively. The former assumes at least some interaction between bedrock and precipitation prior to entering the aquatic regime. This is probably valid for most of eastern Canada where limestones have either been exposed or buried under carbonate-rich tills by the latest glaciation. In the Hudson Bay Lowland (northernmost Ontario and part of northwestern Quebec) organic deposits blanket up to 80% of the carbonate-rich substrate. Large streams, rivers and lakes in this Region intersect mineral soil.

Moderate Potential to Reduce Acidity

Three combinations of soil and bedrock identified as having a moderate potential to reduce acidity best represent the moderate areas illustrated in Figure 1. In order of decreasing coverage they are shallow loam or low lime soils over granite (M3), deep loam or low lime soils (M7), and shallow loam or low lime soils overlying bedrock of moderate potential to reduce acidity (M6). The distribution of each of these three moderate classes is highly variable across eastern Canada. Many of these areas occur as scattered pockets within the low class and are below the limit of resolution at this map scale (Figure 1).

Low Potential to Reduce Acidity

All five categories of combinations identified as having a low potential class in Ontario, Quebec and Newfoundland/Labrador are deep sand (L4). Shallow sands (L3) are frequently found in the more northerly regions, notably in Quebec and Ontario. These two classes are predominantly acid podzols. Areas of high bedrock exposure (L2) are common to shore zones of lakes and northern areas.

Major areas of organic soils overlying noncalcareous bedrock (L5) are identified in western New Brunswick, southern Labrador and Newfoundland and in the west central portion of Quebec. Large areas of peatland are identified adjacent to the Hudson Bay Lowland in northwestern Ontario. Throughout central and western Ontario small pockets of organic soils are common but cannot be discerned at this scale (1:15,000,000).

Table IV. Predominant Classes for the Potential to Reduce Acidity of Atmospheric Deposition in Eastern Canada, by Area (km^2) and Percent (Based on 1:15,000,000 Scale Version of Map)

		Ontario		*Quebec*		*New Brunswick, Nova Scotia, Prince Edward Island*		*Newfoundland and Labrador*		*Total*	
Potential to Reduce Acidity	*Class*	*km^2*	*%*	*km^2*	*%*	*km^2*	*%*	*km^2*	*%*	*km^2*	*%*
High	H1	438,119	41	61,627	4	20,188	15	20,226	5	503,738	16
	H3	64,115	6	77,034	5	13,459	10			125,935	4
Moderate	M1			30,814	2					31,484	1
	M3	32,057	3	30,814	2	21,534	16	210,349	52	314,837	10
	M6	53,429	5	61,627	4					125,935	4
	M7	53,429	5	107,848	7	22,879	17			188,900	6
Low	L1			46,220	3			24,271	6	62,967	2
	L2			200,288	13	8,075	6			220,386	7
	L3	42,743	4	338,950	22	22,879	17	56,632	14	472,255	15
	L4	309,889	29	539,238	35	13,459	10	72,813	18	944,510	30
	L5	74,801	7	46,220	3	12,113	9	20,226	5	157,418	5
Total		1,068,582	100	1,540,680	100	134,586	100	404,517	100	3,148,365	100

DISCUSSION

The map of eastern Canada presents three levels of aquatic sensitivity based on a multifactor interpretation of terrestrial components. At the compilation scale (1:1,000,000) it represents a more detailed regional assessment of aquatic sensitivity than previously defined [4,5].

A comparison of lake alkalinity data [4] and the three sensitivity classes was completed for an area in south-central Ontario (Figure 3) [28]. Based on a χ^2-test, the relationship between the two variables was found to be significant at the 0.001 level. The observed and calculated expected frequencies for each category are presented in Table V. Low alkalinities (<120 μeq-L^{-1}) are associated with areas having a low potential to reduce acidity. In south-central Ontario, these are predominantly areas of shallow sand overlying granitic bedrock (L3). High alkalinities (>600 μeq-L^{-1}) are most frequently associated with areas having a high potential to reduce acidity. These are largely areas of deep clayey material overlying calcareous bedrock.

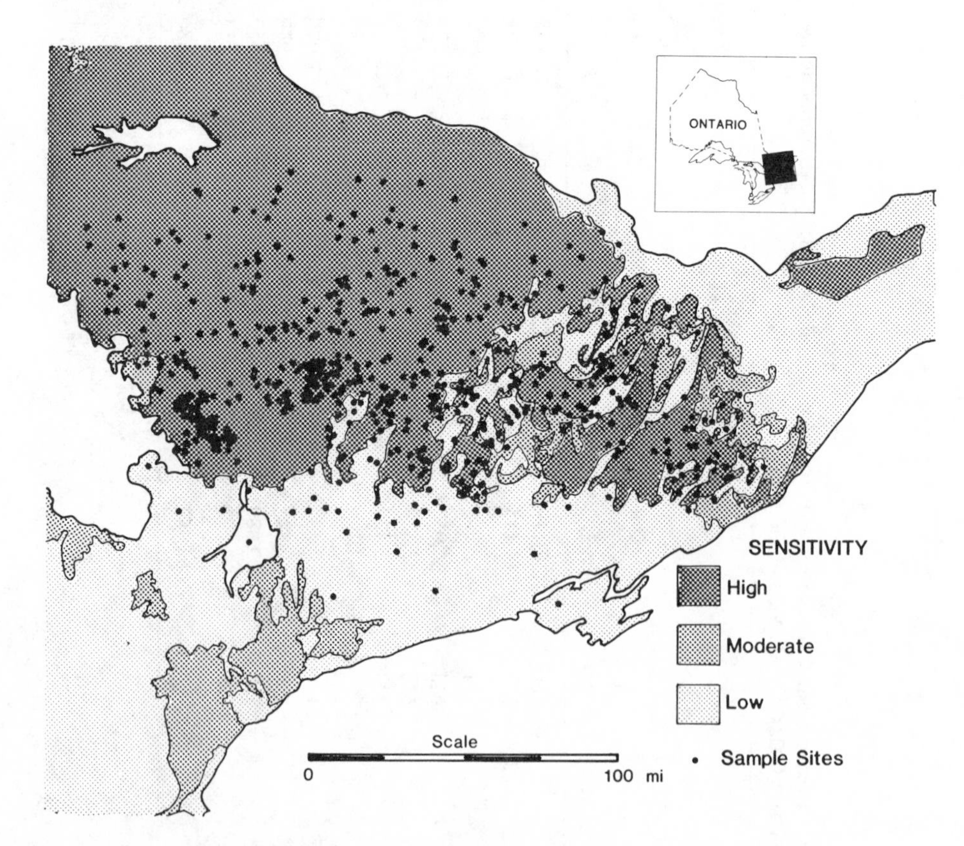

Figure 3. Relationship between measured alkalinities [29] and the potential of soils and bedrock to reduce acidity for south-central Ontario.

Table V. Observed and Expected Frequency of Lakes Alkalinities in Three Sensitivity Levels for South-Central Ontario

Tip Alkalinity ($\mu eq\text{-}L^{-1}$)	*Potential to Reduce Acidity*						*Total Observed*
	Low		*Moderate*		*High*		
	Calculated	*Observed*	*Calculated*	*Observed*	*Calculated*	*Observed*	
<40	59.76	77	4.22	1	15.02	1	79
40–79.9	17.87	94	5.13	2	18.25	0	96
80–119.9	5.35	52	2.78	0	9.89	0	52
120–199.9	36.31	44	2.57	1	9.12	3	48
200–299.9	34.80	37	2.46	2	8.74	7	46
300–599.9	33.28	29	2.35	6	8.36	9	44
600–2000	78.67	43	5.56	11	19.77	50	104
>2000	27.23	6	1.92	4	6.84	26	36
Total		382		27		96	505

The relationship between sensitivity class and alkalinity is best defined where the terrain surrounding sample sites is relatively homogeneous, even at a small scale. Where areas of high and low potential are in close proximity and surface drainage waters cross a number of sensitivity classes, alkalinity values reflect the capacity to reduce acidity of the most reactive soil bedrock category upstream. Thus, local and regional hydrological and hydrogeological conditions need to be assessed when comparing measured aquatic chemistry with potential of areas to reduce acidity. The importance of detailed watershed-based analyses is obvious. Regional assessments are not suitable for close comparison with site specific water chemistry although general relationships can be identified.

Areas having a low potential to reduce the acidity of atmospheric deposition should not be interpreted as representing the total area of already acidified lakes and streams in eastern Canada. These are the areas where acidification would be most pronounced, provided the input of anthropogenic acids (1) add significantly to natural acid production in soils or (2) exceed the buffering capacity of bedrock-dominated areas.

Soils in areas of low potential to reduce acidity are primarily acid podzols (spodosols). Boreal and northern temperate podzols are characterized by the accumulation of organic matter and sesquioxides of iron and aluminum [28]. They contribute natural acidity to surface waters [29]. Thus, the most susceptible areas with respect to the acidification of surface waters are those areas having a low capacity to reduce acidity and are also receiving significant inputs of anthropogenic acids. A loading of 20 $kg\text{-}ha^{-1}\text{-}y^{-1}$ (as SO_4^{2-} in precipitation) has been identified as being "clearly associated with degradation of the more sensitive surface waters" [30]. In eastern Canada approximately 405,000 km^2 have a high sensitivity and estimated annual deposition rates of at least 20 $kg\text{-}ha^{-1}\text{-}y^{-1}$ (Figure 1).

Soils in the most sensitive regions of eastern Canada have low base contents but high sesquioxide and organic matter content. The main cations available to balance increased sulfate in these soils are H^+ and Al^{3+}. Toxic concentrations of Al^{3+} may occur in surface waters draining acid podzols [31]. The capacity of Canadian podzols to adsorb sulfate and thus reduce the mobility of Al^{3+} and anthropogenic acids is not certain. High organic matter content, as found in ferro-humic podzols, is believed to block the sulfate adsorption process [32,33].

CONCLUSIONS

The map and methodology presented here provide an estimate of the extent and distribution of potentially sensitive surface waters in eastern Canada. Within each sensitivity class are several different combinations of soil and bedrock characteristics. Areas identified as having a high potential to reduce acidity account for 20% of eastern Canada. Of this, 75% is associated with limestone bedrock most of which is found in Ontario. A total of 21% of eastern Canada has been identified as having a moderate potential to reduce acidity. The largest moderate class consists of shallow sandy soils overlying bedrock of moderate potential. This particular combination covers more than 60% of Labrador alone.

According to this interpretation, 59% of eastern Canada has a low potential for the reduction of surface water acidity. Large portions of Ontario and Quebec (29 and 35%, respectively) are associated with deep sands or nonlimey soils. These are predominantly acidic podzols and account for approximately 30% of areas identified with a high sensitivity.

Within eastern Canada 25% of the lands receive at least 20 $kg\text{-}ha^{-1}\text{-}y^{-1}$ of wet SO_4^{2-}. Of this area 53% (or approximately 405,000 km^2) is susceptible, being both under a high acidic input and having sensitive waters. These are the areas where surface waters are most likely to be affected by acidic deposition.

Further study is required to improve the resolution of mapping. Both the quality and quantity of available data need to be increased. It is critical that improved soil chemical data, particularly exchangeable base content and sulfate adsorption capacity, be acquired for eastern Canada.

Once the information within the data base has improved, analyses can be conducted at the watershed level. By relating terrestrial characteristics and aquatic chemistry, the present map may be refined.

REFERENCES

1. Bangay, G.E., and C.I. Harris (Co-chairmen). "Memorandum of Intent on Transboundary Air Pollution (MOI). Phase III Working Paper, Impact Assessment Working Group I," (in press).
2. Hendrey, G.R., J.N. Galloway, S.A. Norton, C.L. Schofield, P.W. Shaffer, D.A. Burns and C.F. Powers. "Geological and Hydrochemical Sensitivity of the Eastern United States to Acid Precipitation," U.S. EPA 600/3-80-024 (Corvallis, Oregon: Corvallis Environmental Research Laboratory, 1980).
3. McFee, W.W. "Sensitivity of Soil to Acid Precipitation," U.S. EPA 600/3-80-013 Corvallis Environmental Research Laboratory, (Corvallis, Oregon: 1980).
4. Harvey, H.H., R.C. Pierce, P.J. Dillon, J.P. Kramer and D.M. Whelpdale. "Acidification in the Canadian Environment: Scientific Criterion for an Assessment of the Effects of Acidic Deposition on Aquatic Ecosystems," NRC Report No. 18475 National Research Council of Canada, Ottawa, Ontario (1981).
5. Shilts, W.W., K.D. Card, W.H. Poole and B.V. Sanford. "Sensitivity of Bedrock to Acid Precipitation: Modification by Glacial Processes," GSC Paper 81-14 Geological Survey of Canada, Ottawa, Ontario (1981).
6. Wang, C., and D.R. Coote. "Sensitivity Classification of Agricultural Land to Long-Term Acid Precipitation in Eastern Canada," Contribution No. 98, Land Resource Res. Inst., Agriculture Canada, Ottawa, Ontario (1981).
7. Cowell, D.W., A.E. Lucas and C.D.A. Rubec. "The Development of an Ecological Sensitivity Rating for Acid Precipitation Impact Assessment," Working Paper No. 10, Lands Directorate, Environment Canada, Ottawa, Ontario (1981).
8. "Second Report of the U.S.–Canada Research Consultation Group on Long Range Transport of Air Pollutants," Bilateral Research Consultation Group, Environment Canada, Ottawa, Ontario (1980).
9. Lee, J.J., and D.E. Weber. "Effects of Sulfuric Acid Rain on Two Model Hardwood Forests," U.S. EPA 600/3-80-014 (Corvallis, Oregon: Corvallis Environmental Research Laboratory, 1980).

10. Glass, N.R., D.E. Arnold, J.N. Galloway, G.R. Hendrey, J.J. Lee, W.W. McFee, S.A. Norton, C.F. Powers, D.L. Rambo and C.L. Schofield. "Effects of Acid Precipitation," *Environ. Sci. Technol.* 16(3):162–169 (1982).
11. Kramer, J.R. "Geochemical and Lithological Factors in Acid Precipitation," in *Processes of the 1st International Symposium on Acid Precipitation and the Forest,* L.S. Dochinger, and T.S. Seliga, Eds., General Technical Report NE-23 (Columbus, OH: USDA Forest Service, 1976).
12. Kramer, J.R. "Geochemical Factors and Terrain Response to Environmental Contaminants," Dept. Geology, McMaster University, Hamilton, Ontario (1977).
13. Bobée, B., Y. Grimard, M. Lachance and A. Tessier. "Nature et Etendue de L'acidification des Lacs du Québec," Institut National de la Recherche Scientifique, Ste-Foy, Quebec (1982).
14. Clair, T.A., and D.R. Engstrom. "Sensitivity of Surface Waters of Newfoundland and Labrador to Acidic Precipitation," Technical Report, Inland Waters Directorate, Environment Canada, Moncton, New Brunswick (1982).
15. Clair, T.A., J.P. Witteman and S.H. Whitlow. "Acid Precipitation Sensitivity of Canada's Atlantic Provinces," Technical Report, Inland Waters Directorate, Environment Canada, Moncton, New Brunswick (1982).
16. Kaplan, E., H.C. Thode, Jr. and A. Protas. "Rocks, Soils and Water Quality. Relationships and Implications for Effects of Acid Precipitation on Surface Water in the Northeastern United States," *Environ. Sci. Technol.* 15:539–544 (1981).
17. "Ontario Mineral and Natural Resources," Ontario Geological Map Series, Maps No. 2200 and 2201, Toronto Ontario (1970).
18. "Ecodistrict du Quebec," Lands Directorate, Quebec Region, St. Foy, Quebec (1981).
19. "Ecodistrict Maps and Description for the Atlantic Provinces," Lands Directorate, Halifax, Nova Scotia (1981).
20. "Ontario Land Inventory," Ontario Centre for Remote Sensing, Toronto, Ontario (1977).
21. "Ecodistrict Maps and Description for Ontario," Lands Directorate, Toronto, Ontario (1981).
22. Pala, S., and A.N. Boissonneau. "Surficial Geology Map of Northwestern Ontario, Scale 1:500,000," Ministry Natural Resources, Toronto, Ontario (1979).
23. "Wetlands of Canada," Ecol. Land Class. Ser. 14, Wetlands Working Group, Environment Canada, Ottawa, Ontario (1981).
24. Richards, N.R., J.A. Hansen, W.E.J. Worthy and D.E. Irvine. "A Guide to the Use of Land Information," OIP Publ. 79-2, Ontario Inst. of Pedology, Guelph, Ontario (1979).
25. "Newsletter No. 6," Canada Committee on Ecological (Bio-Physical) Land Classification, Lands Directorate, Environment Canada, Ottawa, Ontario (1979).
26. Clayton, J.S., W.A. Ehrlich, D.B. Cann, J.H. Day and I.B. Marshall. "Soils of Canada," Can. Dept. Agric., Ottawa, Ontario (1977).
27. "The Canadian System of Soil Classification," Publ. 1646, Can. Dept. Agric. Canada Soil Survey Committee, Subcommittee on Soil Classification. Supply and Services Canada, Ottawa, Ontario (1978).
28. Stobbe, P.C. "Characteristics and Genesis of Podzol Soils," in *Soils in Canada,* Royal Soc. Can. Spec. Publ. 3, R.F. Legget, Ed. (Toronto, Ontario: University of Toronto Press, 1968), p. 158.
29. Rosenqvist, I.T. "Alternative Sources for Acidification of River Water in Norway," *Sci. Total Environ.* 10:39–44 (1979).

30. Bangay, G.E., and C.I. Harris (Co-chairmen). "Memorandum of Intent on Transboundary Air Pollution (MOI). Phase II Interim Working Paper, Impact Assessment Working Group I.
31. Ulrich, B., R. Mayer and P.K. Khanna. "Chemical Changes Due to Acid Precipitation in a Loess-Derived Soil in Central Europe," *Soil Sci.* 130:193–199 (1980).
32. Johnson, D.W., and G.S. Henderson. "Sulphate Adsorption and Sulfur Reactions in a Highly Weathered Soil Under a Mixed Deciduous Forest," *Soil Sci.* 128:34–40 (1979).
33. Singh, B.R., G. Abrahamsen and A. Stuanes. "Effect of Simulated Acid Rain on Sulfate Movement in Acid Forest Soils," *Soil Sci. Soc. Am. J.* 44:75–80 (1980).

INDEX